AUTOMOTIVE
EMISSION
CONTROL
AND
TUNE-UP
PROCEDURES

1975-76 Edition

ignition
manufacturers
institute

PRENTICE-HALL, INC., Englewood Cliffs, New Jersey

ISBN: P-0-13-054809–X
ISBN: B-0-13-054825–1

Library of Congress Catalog No.: 75–12614

Printed in the United States of America.

10 9 8 7 6 5 4 3 2 1

PRENTICE-HALL INTERNATIONAL, INC., London
PRENTICE-HALL OF AUSTRALIA, PTY. LTD., Sydney
PRENTICE-HALL OF CANADA, LTD., Toronto
PRENTICE-HALL OF INDIA PRIVATE LTD., New Delhi
PRENTICE-HALL OF JAPAN, INC., Tokyo
PRENTICE-HALL OF SOUTHEAST ASIA (PTE.) LTD., Singapore

TABLE OF CONTENTS

A AC Charging Circuit Indicator Lamps ... 179
AC Charging System (With Ammeter) 171
AC Charging System
  (With Indicator Lamp) 175
AC System Service Precautions 217
Accelerating System 75
Air Injection Pump 327
Alternator and Regulator
  Quick Checks 186
Alternator and Regulator Tests 185
Alternator Charging System 159
Alternator Components 163
Alternator Testing — Chrysler 199, 203
Alternator Testing — Ford 205
Alternator Testing — General
  Motors Delcotron 193, 197
Alternator Testing Factors 180
Assist Units 107
Automatic Choke 79
Automotive Electrical System 31
Automotive Ground Circuits 33
Available Voltage 291

B Battery Capacity Test 25
Battery Leakage Test 19
Battery Rating Methods 28
Battery Specific Gravity Test 21
Battery Three Minute Charge Test 25
Battery Visual Checks 17
Breaker Point Color 348
Breaker Point Wear Patterns 347

C Capacitor Discharge Ignition System ... 307
Carburetor Adjustment — Nonexhaust
  Emission Controlled Engines 379
Carburetor Adjustment — Exhaust
  Emission Controlled Engines 375
Carburetor "Lean Roll"
  Adjustment 375
Catalytic Converter 328
Centrifugal Advance Mechanism 237
Charging Circuit 129
Charging System Resistance Tests 157
Charging Voltage Test 381
Choke 77
Coil Polarity 289
Combination Emission
  Control (CEC) Valve 265
Compression Test 15, 337
Condenser Action 279
Condenser Construction 277
Condenser Testing and Selection 279
Connecting the Timing Light 356
Course Objective 5
Cranking Voltage Test 335
Current Regulator 143

Cutout Relay 139
Cylinder Numbering Sequences
  and Firing Orders 293

D Deceleration Vacuum Advance Valve ... 246
Deceleration Vacuum Advance Valve
  Test 322
Deceleration Valve 117
Deceleration Valve Test 324
Diode 165
Diode-Rectified Output 169
Diode Tests 209, 211
Distributor 343
Distributor Assembly 225
Distributor Cap and Rotor 283
Distributor Advance Solenoid 247
Distributor Advance Solenoid
  Test 323
Distributor Retard Solenoid 247
Distributor Retard Solenoid Test 323
Distributor Vacuum Control Switch 269
Double-Acting Vacuum Spark Control
  Unit 245
Double Contact Voltage Regulator 145
Dual Breaker Points 233
Dual Breaker Point Adjustment 353
Dual Diaphragm EGR Valve 123
Dual Diaphragm Vacuum Spark
  Control Unit 245
Dwell Angle and Variation 229, 352
Dwell Angle Adjustment 351
Dwell Angle Variation Test 352

E Electric Choke Assist Systems 83
Electrical Circuits 49
Electrical Terms 45
Electromagnetic Fields 57
Electronic Distributor Modulator 253
Electronic Ignition System 309
Electronic Spark Control
  (ESC) System 257
Exhaust Emission Control Systems
  Air Injection Type 101
Exhaust Emission Control System
  Assist Units 245
Exhaust Emission Control Systems
  Cleaner Air System (CAS) 97
Exhaust Emission Control Systems
  Engine Modification Types 95
Exhaust Emission Control System Tests
  Air Injection Type 367
Exhaust Emission Control System Tests
  Cleaner Air System (CAS) 363
Exhaust Gas Recirculation Systems 119
Exhaust Gas Recirculation
  Systems Tests 325

F Field Winding Tests 213
Float System 65
Four Stroke Cycle 11
Fuel Evaporation Emission
   Control Systems 125
Fuel Pump 63
Fuel System 59
Fusible Links and Wires 31

G Generator Circuits 147
Generator Function 135
Generator Operating Principles 133
Generator Polarity 149
Generator and Regulator
   Quick Checks 155
Generator and Regulator Tests 151
Generator Testing 137

H Heated Carburetor Air Systems 109
Heated Carburetor Air System Tests .. 371
Hot Idle Compensator Valve 85

I Idle Circuit 67
Idle Limiters 375
Idle Stop Solenoids 113
Ignition Circuit 221
Ignition Coil Action 277
Ignition Coil Construction 277
Ignition Coil Replacement 278
Ignition Manufacturers Institute 3
Ignition System Operation 223
Ignition Timing 355
IM Tune-Up Procedure 331
Imported Car Emission Control
   Dual Intake Manifold System 105
Introduction to Vehicle Emission
   Control Systems 88

M Magnetism 53
Main Metering System 69
Maintenance-Free Battery 26
Manifold Heat Control Valve 87
Merchandising Tune-Up Service 383
Message to the IMI Student 1
Micro-Circuit Voltage Regulators ... 189

O Ohm's Law 47

P Ported Spark Control 239
Positive Crankcase
   Ventilation Systems 91
Positive Crankcase Ventilation
   System Tests 359
Power Loss 13
Power System 73

R Radio Frequency Interference
   Shield 286

Residual Magnetism 55
Resistor Ignition Cables 285
Resistor Spark Plugs 286
Rotor Gap 286

S Secondary Circuit Suppression 285
Side-Pivoted Breaker Plate 230
Spark Advance 243
Spark Advance Timing 235
Spark Delay Valve 261
Spark Plug Features 297
Spark Plug Heat Range 295
Spark Plug Reach 295
Spark Plugs 341
Speed Control Switch (SCS)
   System 263
Starter Ground Circuit Test 43
Starter Insulated Circuit Test 41
Starter Solenoid Test 43
Starting Circuit 37
Starting Motor
   Amperage Draw Test 39
Stator Winding Tests 215
Supplementary Tune-Up Services 333

T Temperature Corrected Hydrometer .. 23
Testing Essentials 9
Testing Exhaust Emission Control
   System Assist Units 321
Testing Sections 7
Thermostatic Vacuum Switch 246
Thermostatic Vacuum Switch Test .. 322
Three Minute Charge Test 25
Transistor Ignition Systems
   Breaker Point Type 301
Transistor Ignition Systems
   Magnetic Pulse Type 305
Transistor Ignition System
   Testing Breaker Point Type 313
Transistor Ignition System
   Testing Magnetic Pulse Type 317
Transistors 167
Transistor Regulators 183
Transmission Controlled Spark
   (TCS) (NOx) 249, 250
Transmission Regulated
   Spark (TRS) System 259

U Underhood Remote Control
   Starting Precaution 338
Unitized Ignition System 311

V Vacuum Advance Mechanism 239
Vacuum Controlled Metering Rod ... 71
Vacuum Spark Control Systems 270
Voltage Regulator 141

W Wiring and Schematic Diagrams 35

CHART INDEX

Chart No.	Title	Page No.
1	The Ignition Manufacturers Institute	2
2	Course Objective	4
3	Testing Sections	6
4	Testing Essentials	8
5	Four Stroke Cycle	10
6	Power Loss	12
7	Compression Test	14
8	Battery Visual Checks	16
9	Battery Leakage Test	18
10	Battery Specific Gravity	20
11	Temperature Corrected Hydrometer	22
12	Battery Capacity Test — Three Minute Charge Test	24
13	Automotive Electrical System	30
14	Automotive Ground Circuits	32
15	Wiring Diagram — Schematic Diagram	34
16	The Starting Circuit	36
17	Starting Motor Amperage Draw Test	38
18	Starter Insulated Circuit Test	40
19	Starter Ground Circuit Test — Solenoid Test	42
20	Commonly Used Electrical Terms	44
21	Ohm's Law	46
22	Electrical Circuits	48
23	Magnetism & Permanent Magnets	52
24	Residual Magnetism	54
25	Electromagnetic Fields	56
26	Fuel System	58
27	Fuel Pump	62
28	Float System	64
29	Idle Circuit	66
30	Main Metering System	68
31	Vacuum Controlled Metering Rod	70
32	Power Jet Operation	72
33	Accelerating System	74
34	Choke Action	76
35	Automatic Choke	78
35A	Electric Choke Assist Systems	82
36	Hot Idle Compensator Valve	84
37	Manifold Heat Control Valve	86
38	Positive Crankcase Ventilation Systems	90
39	Exhaust Emission Control Systems Engine Modification Types	94
40	Exhaust Emission Control Systems Cleaner Air System (CAS)	96
41	Exhaust Emission Control Systems Air Injection Type	100
42	Imported Car Emission Control Dual Intake Manifold System	104
43	Heated Carburetor Air Systems	106
44	Idle Stop Solenoids	110
45	Deceleration Valve	116
46	Exhaust Gas Recirculation Systems	118
46A	Dual Diaphragm EGR Valve	122
47	Fuel Evaporation Emission Control Systems	124
48	The Charging Circuit	128
49	Generator Operating Principles	132
50	Generator Function	134
51	Generator Testing	136
52	Cutout Relay	138
53	Voltage Regulator	140
54	Current Regulator	142
55	Double Contact Voltage Regulator	144
56	Generator Circuits	146
57	Generator Polarity	148
58	Generator and Regulator Tests	150
59	Generator and Regulator Quick Checks	154
60	Charging System Resistance Tests	156
61	Alternator Charging System	158
62	Alternator Components	162
63	Diode	164
64	Transistors	166
65	Diode-Rectified Three Phase Output	168
66	AC Charging System — With Ammeter	170
67	AC Charging System — With Indicator Lamp	174
68	AC Charging Circuit Indicator Lamps	178
69	Transistor Regulators	182

Chart No.	Title	Page No.	Chart No.	Title	Page No.
70	Alternator and Regulator Tests	184	100	Condenser Action	278
71	Alternator and Regulator Quick Checks	186	101	Distributor Cap and Rotor	282
72	Micro-Circuit Voltage Regulators	188	102	Secondary Circuit Suppression	284
73	Alternator Testing — General Motors Delcotron	192	103	Coil Polarity	288
73A	Alternator Testing — General Motors Delcotron — With Integral Regulator	196	104	Available Voltage	290
			105	Cylinder Numbering Sequences	292
			106	Spark Plug Heat Range	294
			107	Spark Plug Features	296
74	Alternator Testing — Chrysler Corporation	198	108	Transistor Ignition Systems Breaker Point Type	300
74A	Alternator Testing — Chrysler Corporation	202	109	Transistor Ignition Systems Magnetic Pulse Type	304
75	Alternator Testing — Ford Motor Company	204	110	Capacitor Discharge Ignition System	306
76	Diode Tests	208	111	Electronic Ignition System	308
76A	The Diode Trio	210	112	Unitized Ignition System	310
77	Field Winding Tests	212	113	Transistor Ignition System Testing Breaker Point Type	312
78	Stator Winding Tests	214	114	Transistor Ignition System Testing Magnetic Pulse Type	316
79	AC System Service Precautions	216	115	Testing Exhaust Emission Control System Assist Units	320
80	The Ignition Circuit	220	115A	Air Injection Pump	326
81	The Ignition System	222	116	IMI Tune-Up Procedure	330
82	Distributor Assembly	224	117	Supplementary Tune-Up Services	332
83	Dwell Angle	228	118	Cranking Voltage Test	334
84	Dual Breaker Points	232	119	Compression Test	336
85	Spark Advance Timing	234	120	Spark Plugs	340
86	Centrifugal Advance Mechanism	236	121	Distributor	342
87	Vacuum Advance Mechanism	238	122	Breaker Point Wear Patterns	346
88	Spark Advance	242	123	Dwell Angle and Variation	350
89	Exhaust Emission Control System Assist Units	244	124	Ignition Timing	354
90	Transmission Controlled Spark (TCS) (NOx)	248	125	Positive Crankcase Ventilation System Tests	358
91	Electronic Distributor Modulator	252	126	Exhaust Emission Control System Tests Cleaner Air System (CAS)	362
92	Electronic Spark Control (ESC) System	256	127	Exhaust Emission Control System Tests Air Injection Type	366
93	Transmission Regulated Spark (TRS) System	258	128	Heated Carburetor Air System Tests	370
94	Spark Delay Valve	260	129	Carburetor Adjustment Exhaust Emission Controlled Engines	374
95	Speed Control Switch (SCS) System	262	130	Carburetor Adjustment Nonexhaust Emission Controlled Engines	378
96	Combination Emission Control (CEC) Valve	264	131	Charging Voltage Test	380
97	Distributor Vacuum Control Switch	268	132	Merchandising Tune-Up Service	382
98	Ignition Coil Construction	272			
99	Condenser Construction	276			

Members of the
IGNITION MANUFACTURERS INSTITUTE

P.O. BOX 1406 • EVANSTON, ILLINOIS 60204

 **CASCO PRODUCTS CORPORATION**
512 Hancock Avenue
Bridgeport, Connecticut 06602

 **P & D MANUFACTURING CO., INC.**
Subsidiary of THE BENDIX CORP.
1217 S. Walnut Street
South Bend, Indiana 46620

THE ECHLIN MANUFACTURING CO.
P.O. Box 472
Branford, Conn. 06405

prestolite. **THE PRESTOLITE COMPANY**
Division of ELTRA Corporation
Toledo, Ohio 43602

ESSEX
automotive parts
ESSEX INTERNATIONAL, INC.
Electro-Mechanical Division
6233 Concord Avenue
Detroit, Michigan 48211

 **BORG WARNER CORP.**
Automotive Parts Division
11045 Gage Avenue
Franklin Park, Illinois 60131

 **FILKO IGNITION DIVISION**
F & B MFG. CO.
5480 N. Northwest Hwy.
Chicago, Illinois 60630

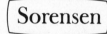 **P. SORENSEN MFG. CO., INC.**
1115 Cleveland Avenue
Glasgow, Kentucky 42141

GUARANTEED PARTS CO., INC.
Auburn Road
Seneca Falls, New York 13148

 **STANDARD MOTOR
PRODUCTS, INC.**
37 - 18 Northern Blvd.
Long Island City,
New York 11101

 **KEM MANUFACTURING CO., INC.**
River Road & Maple Avenue
P.O. Box 351
Fair Lawn, New Jersey 07410

 **TUNGSTEN CONTACT MFG. CO., INC.**
7311 Cottage Avenue
North Bergen, New Jersey 07047

NIEHOFF
AUTOMOTIVE PRODUCTS
C. E. NIEHOFF & CO.
4925 W. Lawrence Avenue
Chicago, Illinois 60630

 WELLS MFG. CORPORATION
2 - 26 S. Brooke Street
Fond du Lac, Wisc. 54935

INTRODUCTION TO CLEAN AIR

The subject of Ecology and particularly that portion of the science that deals with "Clean Air" is receiving national attention. The pollution of the life-sustaining air we breathe is of major concern to everyone.

The automobile has been accused of being one of the contributors to the pollution of the atmosphere. The reason the automobile is a polluter is because it uses gasoline as a fuel. Gasoline is composed of hydrocarbons which are compounds that are inducive to the pollution of air when expelled partially unburned from the automobile engine.

The percentage of unburned hydrocarbons, carbon monoxides and oxides of nitrogen emitted from the automobile are relatively insignificant when measured on a per car basis. But the emissions can actually be measured in tons when multiplied by hundreds of thousands of vehicles all operating in a single conjested area.

When carbon monoxide, oxides of nitrogen and unburned hydrocarbons, all combustion byproducts contained in the gasoline engine exhaust, are emitted into the atmosphere in the presence of sunlight and still air, a haze called photochemical smog is generated. This smog is the lung iritating, eye burning, visibility reducing fog that frequently envolops a metropolitan area with a high vehicle population.

Due to an emergency situation created by photochemical smog in Los Angeles, the State of California enacted legislation in 1961 limiting the emissions from the crankcase ventilation system of all cars sold in their state. Since then California has been the leader in initiating standards to control exhaust emissions and fuel evaporation emissions. Standards enacted in California frequently become Federal standards the year after they are introduced in California.

Other states and cities presently have, or are contemplating, legislation to establish standards to which all vehicles must conform to be eligible for licensing.

TUNE-UP and EMISSION CONTROL

The control of objectionable emissions from the automobile can be achieved and maintained in only one way—by proper engine tune-up. Tune-up and emission control are both performed at the same time by the same operations. By today's standards, performing a professional tune-up using the necessary test equipment and accurate specifications, following the specified test procedures and using quality replacement parts, achieves two desirable results at the same time. The engine is tuned to provide new-car performance, reliability and economy—while vehicle emissions are reduced to conform to specified limits or legislation standards, at the same time.

CAR OWNER'S RESPONSIBILITY

As the smog problem has increased, the nation's car owners have been urged for some years to have their engines tuned at regular intervals as a pollution control measure. Some car owners have responded, others have not.

In a move toward a more compulsory attitude, the Federal Government took action in 1972 requiring the car manufacturers to develop emission control systems and devices that, with nominal maintenance, would effectively control emissions for a period of 5 years or 50,000 miles. This requirement is part of the government's increasing regulation on motor vehicle reliability. Further, the car builder was instructed to inform purchasers of his vehicles of their responsibility to maintain the emission systems on their vehicles so that the emission warranty would stay in effect. Every car builder placed a folder titled "Emission Systems Warranty and Maintenance Schedule" in the glove compartment of every 1972 model vehicle along with the Car Owner's Manual. The folder contains a listing of the services to be performed and the service interval in months or miles for each service.

It is now become the car owner's responsibility to have the recommended services performed at the interval specified, at his own expense. In addition, he must maintain records of the services performed, the date and mileage, and keep receipts. This evidence of emission control system maintenance must be made available on request from the car dealer in the event that emission system trouble occurs. Without such evidence, the car owner's emission warranty may not be honored.

The car owner is obliged to maintain these emission system records for a period of 5 years or 50,000 miles. This requirement is "in addition" to the conventional 12 month or 12,000 mile warranty for all other parts of the vehicle.

TUNE-UP TECHNICIAN'S RESPONSIBILITY

The car owner has met his responsibility when he brings his car into the tune-up bay. Responsibility then rests on the tune-up specialist. He must be qualified to expertly tune the engine and service the emission systems. He qualifies himself by having kept himself abreast of technological advancements in the designs of today's engines.

The tune-up specialist must remember that the Federal Clean Air Act states that it is unlawful for any person to remove or render inoperative any device or element designed into the vehicle's emission control system or units after the vehicle has been sold and delivered to the ultimate user. Violation carries a threat of a $10,000 fine. It is very likely State-sponsored Clean Air Acts will also contain responsibility clauses of a similar nature.

The time is quickly approaching when tuning an engine to emission standards is not only desirable – but mandatory.

This IMI Tune-Up Text Book is emission-control oriented to assist the tune-up technician in acquiring this essential knowledge.

A MESSAGE TO THE IMI STUDENT—

This course has been designed to offer you the sound, basic fundamentals of engine tune-up and emission controls. Your success as a tune-up specialist is governed to a large measure by your understanding of the operating principles of the units and elements involved in a tune-up and in emission control systems and devices you work on.

Take this training opportunity seriously. Apply what you have learned and you will find that because of the understanding that you have gained here, your work has become easier, your trouble-shooting has become more efficient and methodical, and your tune-up operation is producing finer tuned engines and greater customer satisfaction. Above all, you will find that there will be a decided increase in the profits realized from your tune-up effort.

To provide you with a complete understanding of the elements of tune-up, the following subjects are covered in this course. Engine operating fundamentals, batteries, basic electricity, fuel pumps, carburetor circuits, crankcase ventilation systems, exhaust emission control systems and their assist units, evaporation emission control systems, complete coverage of charging systems, complete ignition system coverage, and a step-by-step tune-up procedure. The first portion of the text book covers the theory of each of these subjects. The second portion of the text book covers testing and adjustment procedures along with a step-by-step major tune-up procedure plan.

Every year, increased emphasis is being placed on limiting automotive emissions that contribute to air pollution. The late-model automobile engine is designed from the drawing board up to be a cleaner burning engine than its older cousins. However, the benefits designed into the emission control systems and devices will be realized only as long as the engine is kept properly tuned. This text book is heavily oriented towards coverage of the various emission control systems, and their function, operation, testing and servicing as an essential part of today's tune-up.

Keep your Text Book handy—and refer to it frequently. It contains the information that will refresh your memory, thereby adding to your knowledge and further qualifying you as an automotive tune-up and emission control specialist.

IGNITION MANUFACTURERS INSTITUTE

Date _____

Course sponsored by _____

Distributor Representative _____

Instructor _____

The IGNITION MANUFACTURERS INSTITUTE

The IGNITION MANUFACTURERS INSTITUTE, through the combined resources, knowledge, experience and service abilities of its independent ignition parts manufacturing members, provides the following for the independent automotive parts distributors and their service shop customers:

1. A permanent and profitable business

2. Quality precision-built parts

3. Competitively and profitably priced parts

4. Immediate availability of parts through wide distribution

5. Complete educational program

THE IGNITION MANUFACTURERS INSTITUTE

The Ignition Manufacturers Institute is composed of a group of independent automotive parts manufacturers who have organized to effectively serve the automotive parts distributors, service shops, and the motoring public.

The basic function of the Institute is to assist automotive parts distributors and service shops to operate effectively and profitably by making available quality precision-manufactured parts. This, in turn, allows them to offer their customers prompt and satisfactory service.

The Ignition Manufacturers Institute, through the combined resources, knowledge, experience, and service abilities of its members provides the following benefits for parts distributors and their service shop customers:

1. A permanent and profitable business.

2. Quality, precision-built parts.

3. Competitively and profitably priced parts.

4. Immediate availability of parts through wide distribution.

5. Educational programs.

The Ignition Manufacturers Institute and its members have the sincere desire to make available premium quality automotive parts, constant service, technical publications, service manuals, and an automotive tune-up course — all of which have been developed to assist the automotive serviceman to operate a successful and profitable business and to help raise his vocational standard to the level of other professional tradesmen.

COURSE OBJECTIVE
The objective of this Ignition Manufacturers Institute's Tune-Up Training Course is:

- To train better mechanics who give better service

- To build a reputation as a skilled tune-up specialist

- To develop more satisfied customers

- To make more profit

4

COURSE OBJECTIVES

These are the major objectives of the Ignition Manufacturers Institute Tune-Up Course:

TO TRAIN BETTER MECHANICS WHO GIVE BETTER SERVICE. Because of the training a mechanic receives, he develops a deeper understanding of engine operating principles and the fundamentals of the electrical systems that are included in a tune-up procedure. He can, therefore, locate trouble quickly and correct it effectively. This is a highly important factor in providing good service.

TO BUILD A REPUTATION AS A SKILLED TUNE-UP SPECIALIST. One of the finest reputations an automotive serviceman can have is being known as a man who is skilled in diagnosing trouble and in quickly restoring the engine to its original performance and bringing it within emission limits. Since engine tune-up is one of the highest profit phases of automotive service and emission control standards are being instituted in many states, a specialist's reputation in this field is invariably associated with a better than average income.

TO DEVELOP MORE SATISFIED CUSTOMERS. A skilled man equipped with the knowledge of his trade, the necessary pieces of testing equipment and a sound understanding of test procedures is in the best possible position to develop satisfied customers. Finding trouble quickly and expertly, and correcting it properly, is the surest way of satisfying the car owner.

TO PROVIDE BETTER SERVICE AT A PROFIT. Because the trained man has the understanding and the ability to quickly locate trouble, he can handle more work in a given period of time. He uses test equipment intelligently, knowing that its use is essential to providing quality tune-up. The combination of his knowledge and the use of his equipment increases the sales volume of the quality ignition parts he handles and it also increases the services he has to sell. The result is reflected in increased profits from his tune-up effort.

TESTING SECTIONS

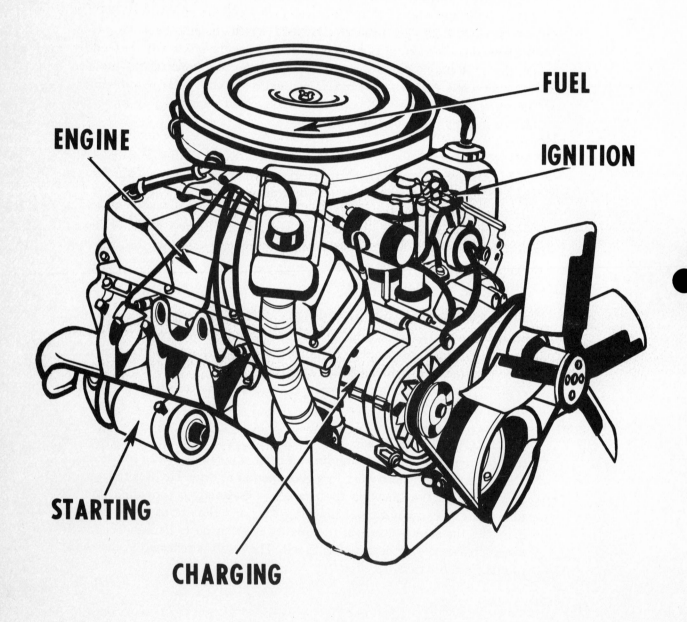

FUEL

ENGINE

IGNITION

STARTING

CHARGING

TESTING SECTIONS

There are five important engine areas that must function properly in order to insure good performance, economy and dependability — the three factors in which practically every motorist is interested. The five areas to be considered are: the engine, the starting system, the charging system, the ignition system and the fuel system. Only when these five areas are restored to normal operating condition can a satisfied customer be developed and greater profits realized. These five areas, as they influence tune-up, will be covered in this course.

The emission control systems and assist devices designed into the modern automobile are NOT systems separate from the five testing areas. These systems and units are part of the distributor, ignition system, ignition timing, carburetor and fuel system and must be tested and serviced at the same time the basic system is being diagnosed and tuned. This is the only way the desired results of a tuned engine and limited emissions can both be achieved at the same time.

Much of this course will be devoted to a study of the electrical system due to the great part it plays in the over-all function and performance of the automobile. Almost every important part or operation of the automobile depends to some degree on the proper functioning of the electrical system.

The electrical accessories on the late model automobiles are many in number compared to the early vehicles. Electrical systems on today's vehicles are being called upon to perform functions beyond those required a few years ago. Although service operations on the modern automobile sometimes seem difficult, the proper understanding of the units in the electrical system and their function convert these tasks into quickly accomplished profitable service jobs.

This course has been developed to assist you by increasing your understanding of these tune-up service operations.

TESTING ESSENTIALS
The five elements of quality automotive tune-up are:

1 - Trained men

2 - Dependable test equipment

3 - Accurate specifications

4 - Specific test procedure

5 - Quality replacement parts

TESTING ESSENTIALS

There are FIVE basic elements needed to perform a quality tune-up.

THE FIRST ESSENTIAL IS A TRAINED MAN. The success of a tune-up business is particularly dependent on this factor. A trained man is an observant serviceman who by virtue of his knowledge, ability and experience is able to locate the trouble and to restore the vehicle to its original operating condition quickly and efficiently. Since he is a trained individual, he knows the value of, and gets the maximum results from, the test equipment he uses. This is true because test equipment is only as good as the man who operates it.

THE SECOND ESSENTIAL IS DEPENDABLE TEST EQUIPMENT. The use of dependable test equipment is an extension of the serviceman's senses permitting him to locate trouble quickly and accurately. The increasing complexity of the modern automobile engine makes the use of test equipment a necessity. Several members of the Ignition Manufacturers Institute merchandise a line of dependable test equipment.

THE THIRD ESSENTIAL IS ACCURATE SPECIFICATIONS. Accurate specifications serve as a standard to determine the service tolerances of the vehicle. By comparing the readings obtained from the test equipment with the specifications, the trained serviceman can quickly determine the service necessary to restore the vehicle to an efficient operating condition. These specifications are compiled by your Ignition Manufacturers Institute supplier.

THE FOURTH ESSENTIAL IS A SPECIFIC TEST PROCEDURE. The test procedure specified in this course contains the elements necessary to cover the greatest majority of engine and accessory malfunctions. Further, the steps of this procedure are arranged in a sequence that permit the tests to be easily conducted, in a logical order, with a minimum expenditure of time.

THE FIFTH ESSENTIAL, EQUAL IN IMPORTANCE TO ANY OTHER, IS QUALITY REPLACEMENT PARTS. The best assurance a tune-up specialist has of performing a tune-up that will keep an engine operating efficiently for thousands of miles is the use of quality replacement parts. The Ignition Manufacturers Institute organization sponsoring this training course, features premium quality parts.

Remember — your tune-up business can only be as good as the service you render and the products you use.

FOUR STROKE CYCLE

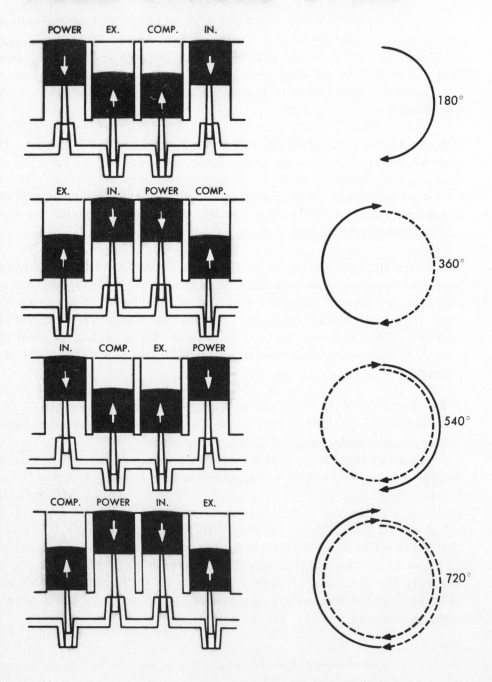

FOUR-STROKE CYCLE

The gasoline engine is an internal combustion heat engine that develops its power by burning a mixture of gasoline and air in the engine's cylinders. This engine is designed to operate on a 4-stroke cycle principle. Each stroke is a movement of a piston, either upward or downward, in its cylinder.

The strokes occur in the following sequence: intake, compression, power, and exhaust.

The intake stroke (downward) serves to draw the air-fuel mixture from the carburetor into the cylinder through an open intake valve. The compression stroke (upward) compresses the mixture to approximately 150 pounds per square inch, a pressure at which the gaseous mixture is suitable for efficient combustion. The power stroke (downward) is the result of the spark plug igniting the compressed fuel charge. The rapid burning of the fuel produces tremendous heat and pressure which expands the gases and raises the combustion pressure to more than four times the compression pressure. This pressure, exerted on the head of the piston, is the force that produces the power that drives the vehicle. The exhaust stroke (upward) forces the burned gases out of the cylinder through the open exhaust valve, into the exhaust system. With the cylinder cleansed of the exhaust gases, the cylinder is ready for another intake stroke and a repetition of the 4-stroke cycle. This cycle of strokes, in the sequence listed, continues as long as the engine is in operation.

Theoretically, each stroke lasts for 180 degrees of crankshaft rotation. In actual practice, however, the length of these strokes is modified somewhat for better engine performance and efficiency.

All the cylinders in an engine are fired in every 720 degrees of crankshaft rotation regardless of the number of cylinders in the engine. The more cylinders an engine has, the more power strokes there will be in these two crankshaft revolutions increasing the engine's power output and performance.

POWER LOSS

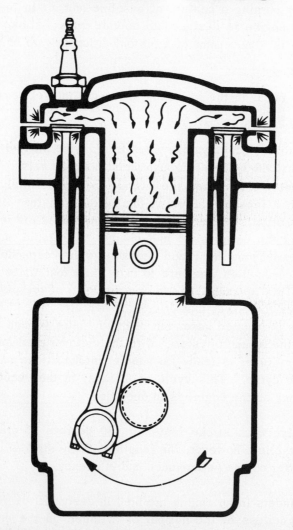

COMPRESSION STROKE LEAKAGE

POWER LOSS

Before a successful tune-up can be performed, it must be determined if the engine is in a satisfactory mechanical condition. An engine that has burned or leaking valves, worn piston rings, sticking valve lifters, leaking cylinder head gaskets or other mechanical malfunctions, will not perform efficiently even after being tuned-up. These conditions **must** be corrected before an engine can be tuned to perform satisfactorily or before emissions can be effectively limited.

Since the power developed by an engine on its power stroke is largely dependent on the efficiency of the compression stroke, and because of the testing convenience afforded, the compression stroke is used for testing engine condition.

During the compression stroke the air-fuel mixture is compressed in the tightly sealed combustion chamber. Should any openings be created by burned valves, leaking gaskets, or worn piston rings, the reduced amount of the air-fuel mixture would proportionately reduce the power output of the engine.

Leakage at any point in the combustion chamber will affect efficient engine operation. Leaking intake valves will allow a portion of the air-fuel mixture to be pushed back into the intake manifold during the compression stroke and less fuel will be available for the power stroke. During the power stroke, the expanding gases will leak past the burned valves and less pressure will be available on the head of the piston. Also burned gases will be forced into the intake manifold to mix with the air-fuel mixture. A diluted air-fuel mixture will then be available for the next intake stroke and consequently less power will be developed by the engine. If the exhaust valve is burned, the expanding gases will leak through it and less power will be available from the cylinder.

Any leakage past the piston rings will also affect the power of the engine. During the compression stroke, part of the air-fuel mixture will be forced into the crankcase and cause oil contamination. The power stroke will also force burned gases into the crankcase. These gases will overheat some of the oil, turning it to carbon and the oil will become contaminated.

A leaking head gasket will permit water to be drawn into the cylinder during the intake stroke. During the compression and power strokes, gases will be forced from the combustion chamber into the cooling system and cause the engine to overheat. Also, a less dense air-fuel mixture will be available for the power stroke.

It is obvious that conditions of compression stroke leakage are proportionately reflected in engine power loss and must be corrected before an engine can be properly tuned.

COMPRESSION TEST

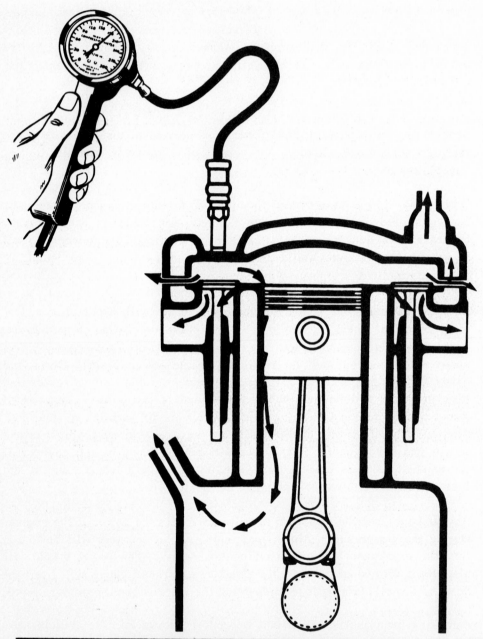

COMPRESSION TEST RESULTS						
Cylinder	1	2	3	4	5	6
Pressure (psi)	130	160	155	157	152	158

COMPRESSION TEST

A compression test is conducted to determine if the piston can compress the air-fuel mixture in the cylinder to a specified pressure which is essential for efficient combustion and maximum power output.

The compression test results are compared to the specifications. The compression pressure specifications will be listed, for example, as 160 pounds per square inch (psi), plus or minus 10 percent:

$$160 \text{ psi} \pm 10\%$$

The plus or minus tolerance is an important part of the specification since it limits the permissible variation between cylinder pressures. Only when ALL the cylinder pressure readings are within the limits of the specification, is smooth engine operation possible.

The figures in the chart table indicate an abnormally low reading in cylinder No. 1. A ten percent reduction in the reading of cylinder No. 2 (160 – 16 = 144) reveals that there is more than the permissible pressure variation between cylinders No. 1 and No. 2. The cause of the low pressure in cylinder No. 1 must be corrected before an effective tune-up can be performed.

If the compression builds up quickly and evenly to the specified pressure on each cylinder and does not vary more than the allowable tolerance, the readings are normal. The engine can be considered acceptable for tune-up.

Worn piston rings will be indicated by low compression on the first stroke which tends to gradually build up on the following strokes. A further indication is an improvement of the cylinder reading when about a tablespoon of motor oil is added to the cylinder through the spark plug hole with an oil can.

Valve trouble is indicated by a low-compression reading on the first stroke and does not rapidly build up pressure with succeeding strokes. The addition of oil will not materially affect the readings obtained.

Leaky head gaskets on two adjacent cylinders will produce the same test results as valve trouble. An additional indication of this particular trouble is the appearance of water in the crankcase.

Carbon deposits result in compression pressures being considerably higher than specified. It is possible that carbon can hide a defect in the cylinder, as the deposit will raise the compression pressure of a cylinder to the extent which might compensate for leakage.

The procedure for conducting a compression test is covered in detail in the Tune-Up Procedure Section of this course.

BATTERY VISUAL CHECKS

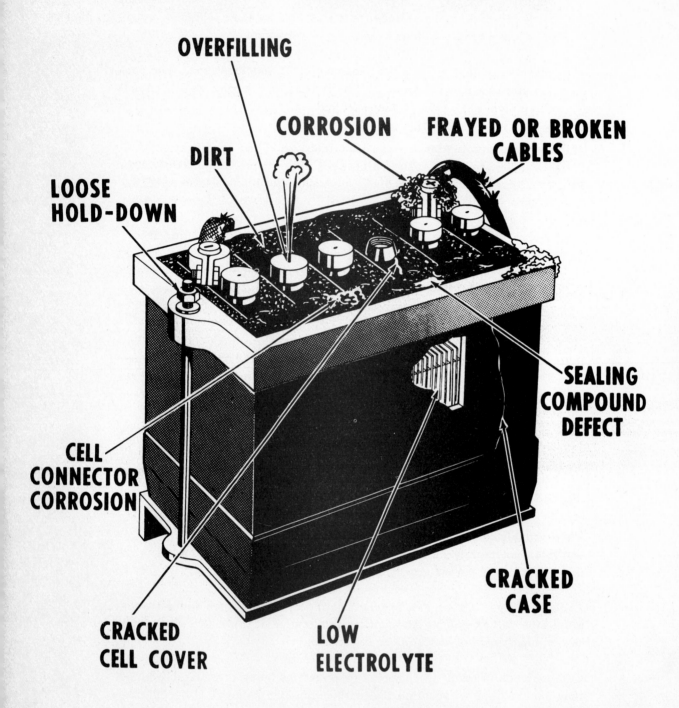

OVERFILLING

CORROSION

FRAYED OR BROKEN CABLES

DIRT

LOOSE HOLD-DOWN

SEALING COMPOUND DEFECT

CELL CONNECTOR CORROSION

CRACKED CASE

CRACKED CELL COVER

LOW ELECTROLYTE

BATTERY VISUAL CHECKS

It has often been stated that the man who tests the most batteries, sells the most batteries.

Battery care and testing are relatively simple. A basic knowledge of how a battery is constructed, how it works, along with a few pieces of test equipment and simple test procedures, will provide any serviceman with the essentials he needs to provide excellent service in this money-making phase of tune-up.

The primary function of the battery is to provide power to operate the starting motor. It must also supply the ignition current during the starting period and accomplish this even under adverse conditions of temperature and other factors.

The battery can also serve, for a limited time, as a source of current to satisfy the electrical demands of the vehicle which are in excess of the output of the generator.

Batteries used in automobiles are known as storage batteries. This term storage battery is sometimes misinterpreted. A battery does not store electricity but does store energy in a chemical form. It accomplishes its task by a chemical process which takes place inside a battery when it is connected to a complete circuit.

Basically stated, a battery is composed of two dissimilar metals in the presence of an acid. The battery is constructed of a series of positive and negative plates. These plates are insulated from each other by means of separators. All the positive plates are interconnected and all the negative plates are interconnected. These interconnected series of positive and negative plates are submerged in the container filled with a sulphuric acid and water solution known as "electrolyte".

The first test of a battery is a visual inspection. If a battery is cracked or otherwise defective, it must be discarded. If the electrolyte level in the battery is low, or if the ground connections or insulated connections are defective, the battery cannot operate efficiently. It is also very important that the battery be kept clean. Dirt and moisture can serve as a conductor and slowly discharge the battery over a period of time.

When activating a dry-charge battery, follow the battery manufacturer's service procedure and be sure to fill each cell properly with the electrolyte supplied. Apply a warm-up charge of 15 amperes for 10 minutes after activating the battery, when so instructed. Observe charging cautions.

Dry-charge batteries will be damaged if moisture enters the battery while it is stored. Always store dry-charge batteries in the coolest, and driest location possible.

BATTERY LEAKAGE TEST

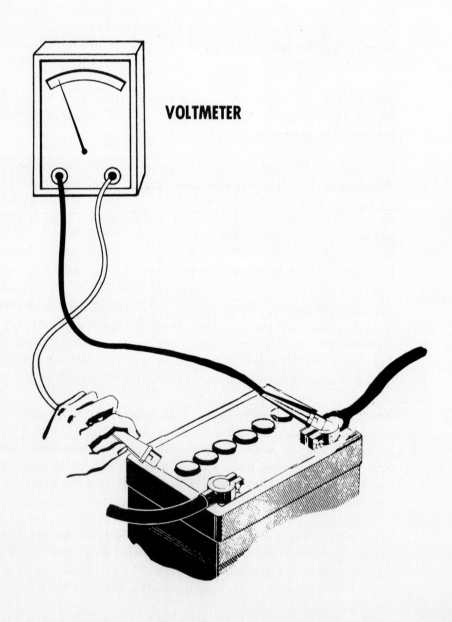

VOLTMETER

BATTERY LEAKAGE TEST

By using a voltmeter, a tune-up specialist can show his customer that a battery with a dirty top is actually leaking current and may become self-discharged.

A leakage test is performed by clipping the negative voltmeter lead to the battery negative terminal. The positive voltmeter lead should be moved over the insulated surface of the top of the battery. Any reading is an indication of electrical leakage. This undesirable electrical path is composed largely of electrolyte which has been expelled from the battery through the fill cap vents by the charging action of the generator. Electrolyte is, of course, a conductor of electricity.

If meter readings indicate any electrical leakage, the battery top should be washed with a solution composed of one spoonful of baking soda mixed in a pint of water. After the bubbling action, induced by acid neutralization stops, rinse the battery top with clean water and dry the battery.

Corrosion accumulation around the cable clamps should be removed with a brush and washed with the same solution.

BATTERY SPECIFIC GRAVITY

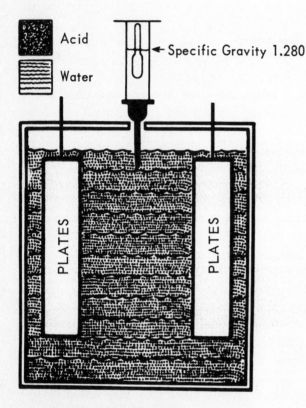

Acid

Water

← Specific Gravity 1.280

FULLY CHARGED

Acid in water gives electrolyte
specific gravity of 1.280

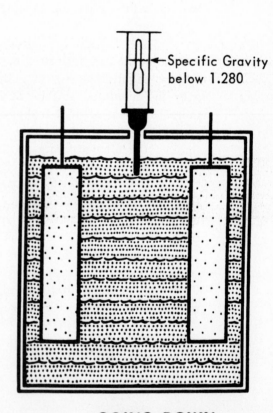

← Specific Gravity
below 1.280

GOING DOWN

As battery discharges, acid begins to
lodge in plates. Specific gravity drops.

BATTERY SPECIFIC GRAVITY TEST

By using a hydrometer, the specific gravity of the electrolyte solution in a battery can be determined. The battery specific gravity is an indication of the battery state of charge. If the state of charge is low, the hydrometer will read low. If the state of charge is high, the hydrometer will read high. As an example, a reading from 1.260 to 1.280 indicates a fully-charged battery. A reading from 1.200 to 1.220 indicates a battery that is in a half-charged condition. Readings below 1.200 indicate a battery is in a discharged condition and cannot give satisfactory service.

The definition of specific gravity is the weight of a liquid compared to the weight of an equal volume of water. The specific gravity of chemically pure water at 80° F. is "one". Therefore, by knowing the specific gravity of sulphuric acid, we can accurately measure the ratio of sulphuric acid to water in the battery electrolyte solution.

When a battery is in a full state of charge, the negative plates are basically sponge lead, the positive plates are lead peroxide, and the electrolyte has a maximum acid content and a minimum water content.

As the battery is discharging, the chemical action taking place reduces the acid content in the electrolyte and increases the water content, while both negative and positive plates are gradually changing to lead sulphate.

When the battery is in a state of discharge, the electrolyte is very weak since it now has minimum acid content and a maximum water content and both plates are predominately lead sulphate. The battery now ceases to function because the plates are now basically two similar metals in the presence of water instead to two dissimilar metals in the presence of an acid.

During the charging process the chemical action that occurred during the battery discharge, is reversed. The lead sulphate on the plates is gradually decomposed, changing the negative plates back to sponge lead and the positive plates back to lead peroxide. The acid is redeposited in the electrolyte returning it to full strength. The battery is now again capable of performing all its functions.

After activating a dry-charge battery, check the specific gravity. The gravity reading should be 1.260 or slightly higher. If the electrolyte level drops shortly after the initial fill, due to the plates and separators absorbing some of the solution, add more electrolyte to bring the solution up to the proper level. When so instructed, charge the battery at 15 amperes for 10 minutes before installing the battery to assure a full charge.

TEMPERATURE CORRECTED HYDROMETER

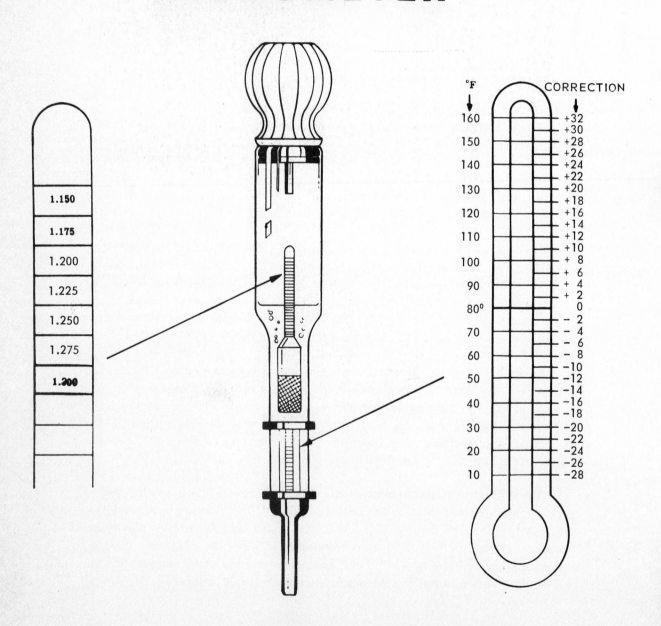

TEMPERATURE CORRECTED HYDROMETER

Hydrometer floats are calibrated to indicate correctly only at 80°F. temperature. If used at any other temperature, a correction factor must be applied. The reason for this lies in the fact that a liquid expands when it is heated and shrinks when it is cooled. This will cause a change in the density of the electrolyte solution which will raise or lower the specific gravity reading.

A thermometer is built into the temperature compensating-type hydrometer. The scale of this thermometer indicates the temperature of the solution. This reading should be used so that the proper temperature correction factor can be applied.

The table is based on an electrolyte temperature of 80°F. For other temperatures, correct the indicated reading by adding 4 points (.004) for each 10° above 80°F. and subtracting 4 points for each 10° that the electrolyte temperature is below 80°F.

For example: A specific gravity reading of 1.230 is obtained at a solution temperature of 10°F. If the electrolyte temperature is disregarded, the reading of 1.230 may be considered as low but acceptable. When the reading is temperature corrected, the true reading of 1.202 (7 × 4 = 28 from 1.230) reveals that the battery is actually very low and definitely in need of charging.

A specific gravity reading of 1.235 is obtained at a solution temperature of 120°F. The reading itself may be interpreted as being rather low but when temperature corrected the reading is actually 1.251 (4 × 4 = 16 added to 1.235). This specific gravity may be high enough for the battery to be restored to full charge by the car's generator.

These examples indicate the importance of temperature correcting specific gravity readings to accurately interpret the true state of battery charge.

To accurately test the true condition of the battery, a light load test or a capacity test should be conducted after the specific gravity has been tested.

12 to 1280 ok
below 1200 new battery

BATTERY CAPACITY (LOAD) TEST

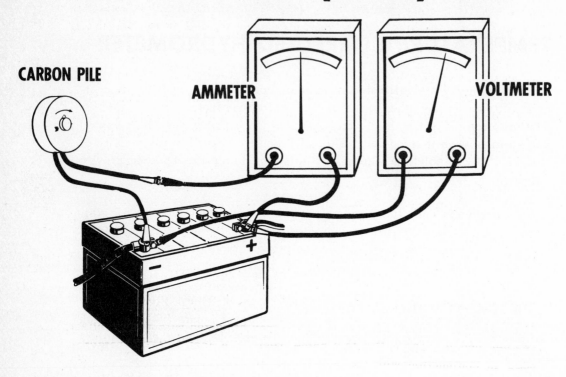

CARBON PILE

AMMETER

VOLTMETER

THREE MINUTE CHARGE TEST

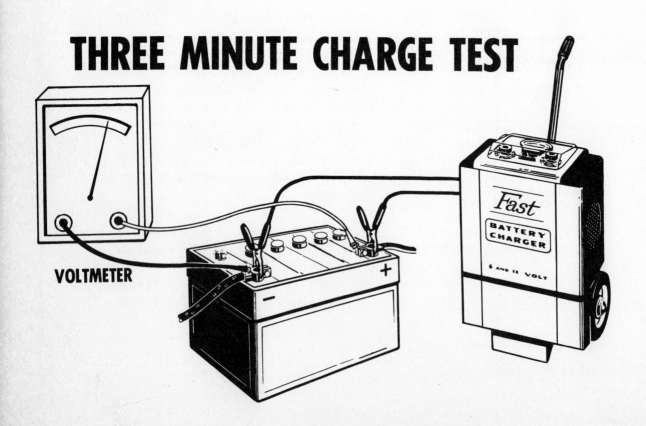

VOLTMETER

Fast

BATTERY CHARGER

6 AND 12 VOLT

BATTERY CAPACITY TEST

Most engine starting failures are caused by the inability of a battery to maintain a voltage high enough to provide effective ignition while cranking a cold engine.

BATTERY CAPACITY TEST

The function of the battery capacity test is to duplicate the battery drain of a cold engine start while observing the battery's ability to maintain voltage. A battery that passes the capacity test will provide dependable performance.

The Battery/Starter Tester has an ammeter, a voltmeter and a carbon pile, which is a battery loading device. The charged battery is discharged at a rate of three times its ampere-hour rating for 15 seconds while its voltage is observed. The voltage of a 12-volt battery should not drop below 9.0 volts or that of a 6-volt battery below 4.5 volts. A reading below this specification indicates a defective battery that should be replaced.

180 - 6v
200 - 12v

THREE MINUTE CHARGE TEST

A battery that is less than fully charged may be tested with a fast battery charger and a "Three Minute Charge Test." A fast battery charger is used in conjunction with the Battery/Starter Tester for this test. Fast charge the battery three minutes at not more than 40 amps for a 12-volt battery and 75 amps for a 6-volt battery. With the charger in operation, observe the voltage of the battery. If the voltage exceeds 15.5 volts for a 12-volt battery, or 7.75 volts for a 6-volt battery, the battery is sulphated or worn out. This indicates that the plates will no longer accept a charge under normal conditions and the battery should be discarded.

BE SURE to observe all precautions relative to working around a battery while it is being charged. Explosive hydrogen gas is liberated from the electrolyte while the battery is being charged. Sparks from a lighted cigarette or from charger clamps being disconnected while current is still flowing may cause an explosion that will destroy the battery and possibly inflict personal injury.

Also BE SURE that all the precautions that are relative to working on a battery installed in a car that is equipped with an alternator are observed. These precautions will be fully covered in our discussion on alternator charging systems. These precautions also pertain to a car equipped with a transistorized ignition system or transistorized fuel injection.

A battery with a one-piece, hard-top cover must be tested with the capacity test since individual cell tests cannot be conducted. Individual cell tests require that the meter test prods contact the cell connectors, by piercing the sealing compound if necessary. Under no circumstance should there be an attempt to pierce the one-piece battery cover with meter prods to conduct an individual cell test.

THE MAINTENANCE-FREE BATTERY

A new type of battery, called "maintenance-free," will be seen on some late model cars. It can easily be identified by its lack of filler caps; the top of the battery is a solid cover. These batteries are designed to operate without periodic additions of water throughout their normal service life. Since the addition of water is the only maintenance normally performed on batteries, the elimination of this service results in a truly maintenance-free battery.

Such batteries are still classed as lead-acid batteries and still function in the same manner as the conventional variety. The only difference is in the lead alloy used to make the plates. Conventional batteries use a lead-antimony alloy for plate construction because antimony increases the strength and casting qualities of lead. Pure lead, by itself, is not a suitable plate material.

Although antimony adds the necessary mechanical properties to lead, it also affects the electrical properties of the battery. It tends to increase a battery's normal self-discharge rate and also to lower its "gassing" potential.

Self-discharge is what causes an otherwise normal battery to gradually run down if left without charging for long periods, especially in temperatures over 60°F. Normally, this is not a problem for batteries in regular service, but can be for new batteries left in stock or for vehicles left unattended for long periods. Eliminating antimony minimizes this internal self-discharge.

"Gassing" is what happens to the electrolyte when a battery is under normal charge. This is what causes the water in the electrolyte to gradually "boil away." Actually, the water does not boil but, instead, is broken down by the charging voltage into its two elements—hydrogen and oxygen. By eliminating antimony from the plates, the voltage at which water breaks down, or "gasses," is raised. Therefore, at normal charging voltages, the original water in the electrolyte of a maintenance-free battery can last several years or more before it finally boils away.

In place of antimony, the lead in a maintenance-free battery is alloyed with calcium to give it the necessary manufacturing characteristics. Calcium, however, is a more difficult material to process and results in higher battery costs. But the result is a battery with a very low self-discharge rate and a greatly reduced tendency to boil off its water.

Although such batteries do not require servicing as do conventional ones, there are times when they must be tested to see if they are still serviceable. This is most commonly done with a battery capacity test as described previously. If the battery voltage under load remains above the specified minimum (typically 9.6 volts), the battery is considered serviceable. However, if it falls below the minimum, it could be either defective *or* merely in a discharged state. Before it can be condemned, further steps are needed.

If the battery has a built-in "hydrometer," evidenced by a small window in the cover, check to see if a green dot appears in the window. If it does, it means that the battery is sufficiently charged (over 1.225 sp. gr.) to be load tested. If it now fails the load test, it means that the battery is defective. However, if it fails the load test without the presence of the green dot, the battery must be recharged until the dot appears. It is then given a second load test.

Batteries without a built-in hydrometer, if they fail the load test, must be fast charged for a period of time specified by the manufacturer. If they fail the load test after the specified charge period, the battery is considered defective.

BATTERY RATING METHODS

Over the years, many methods have been devised to specify the capacity or electrical size of batteries. Presently, only three methods are commonly used: (1) the amp-hour method, (2) cold cranking performance and (3) reserve capacity. In addition to electrical rating methods, batteries are also arranged according to their physical size by "group numbers." Batteries with the same group number have the same dimensions and are physically interchangeable. However, they may have widely varying electrical capacities and for this reason are not always interchangeable.

THE AMP-HOUR METHOD

The Amp-Hour Method has been used for many years, although it is being gradually replaced by the Cold Cranking and Reserve Capacity ratings. A battery's amp-hour rating is determined by discharging a fresh, fully charged battery at a constant rate so selected that at the end of 20 hours the voltage will have fallen to 1.75 volts per cell (or 10.5 volts for a 12-volt battery). This discharge current, times 20 hours, gives the battery's amp-hour rating. For example, if the required discharge current was 3.0 amperes, the battery would be rated at 3 × 20 or 60 amp-hours. It should be noted that this does *not* mean that such a battery can be discharged at 60 amperes for one hour, or any other combination, with the same results. When replacing batteries, always replace with the specified, or higher, amp-hour battery.

COLD CRANKING PERFORMANCE

This is a more recent rating method designed to show a battery's cold weather cranking ability. A Cold Cranking rating shows how many *amperes* can be drawn from a battery at 0°F for 30 seconds before its voltage drops below 1.2 volts per cell (or 7.2 volts for a 12-volt battery). As a rough rule-of-thumb, a battery's Cold Cranking rating in amperes should approximate the engine's displacement in cubic inches. Most new batteries have this rating, and sometimes the amp-hour rating, imprinted on the battery cover. There is no convenient way to convert between amp-hour ratings and cold cranking ratings.

RESERVE CAPACITY

The Reserve Capacity rating is specified in *minutes* and indicates for how long a vehicle can be driven with battery only in the event of charging system failure. The rating is established by noting how long it takes a fully charged battery (at 80°F) to drop to 1.75 volts per cell (or 10.5 volts for a 12-volt battery) at a constant 25 ampere discharge rate. This discharge rate represents a typical nighttime electrical load with headlights and heater. Thus a battery with an 80 minute Reserve Capacity rating could keep a vehicle, with a defective charging system, running for at least one hour and 20 minutes. Operating time without the headlights and heater, of course, would be even longer. It must be remembered, however, that this rating assumes the battery to be fully charged initially. A partly charged battery may not provide the full specified operating time.

AUTOMOTIVE ELECTRICAL SYSTEM

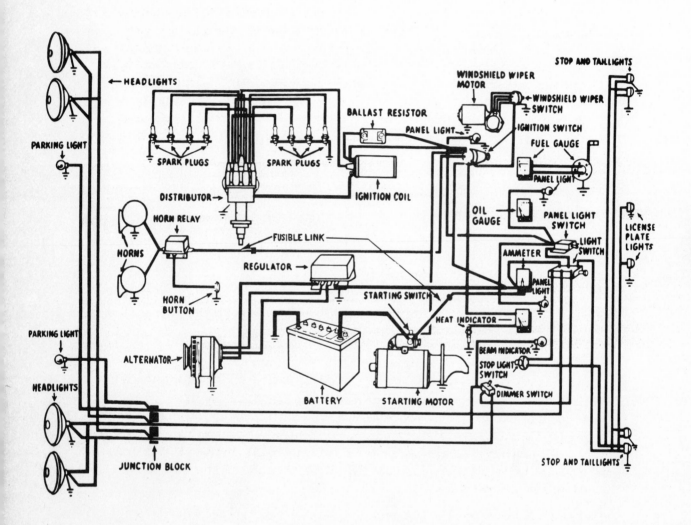

AUTOMOTIVE ELECTRICAL SYSTEM

Pictured on this chart is the charging system, starting system, the ignition system, the lighting system, and various accessory circuits. Today's modern automobile, with its high-compression engine and the addition of various accessories, has brought about an electrical system which has become increasingly complex. With the need for improved starting efficiency, greater demands on the ignition system, and larger generator capacity, new electrical units have been developed and others redesigned. Every unit of the automotive electrical system whether it be a generator, a coil, a solenoid, or a voltage regulator, contains an electrical circuit and depends upon electricity in some form to do its work. The only way to determine the condition of these circuits is by accurately and efficiently testing their components.

4 guages smaller then wire protecting

FUSIBLE LINKS AND WIRES

The function of the fusible link is to protect circuits that are not normally fused, as: charging circuit, ammeter, parking brake alarm, turn indicator circuit, cigarette lighter and others.

One form of fusible link is inserted in the wiring harness usually at a junction block in the engine compartment. Another form attaches the link to the battery insulated cable.

The fusible link is constructed of a fuse wire covered with a Hypalon insulation. The fuse wire is normally four wire gauges smaller than the circuit it is protecting. When the circuit is subjected to a short-circuit overload, the fusible link overheats and burns out protecting instruments and wiring from damage. The Hypalon insulation, which is capable of withstanding extreme temperature, swells to about twice its normal size and assumes a "bubble" appearance. This indicates that the fuse link it contains has burned out.

Burned links are replaced by being cut out of the circuit and replaced with new links of the proper size after the cause of the trouble has been located and corrected.

Another form of this device is a fusible wire used on the charging system voltage regulator on Chrysler Corporation vehicles. If the alternator field current draw should become excessive, as in the event of short circuiting during charging systems tests or accidental grounding of the alternator field lead, the fusible wires will melt thereby protecting the regulator, the alternator and the wiring harness. These wires are also replacable after the cause of the overload is corrected.

AUTOMOTIVE GROUND CIRCUITS

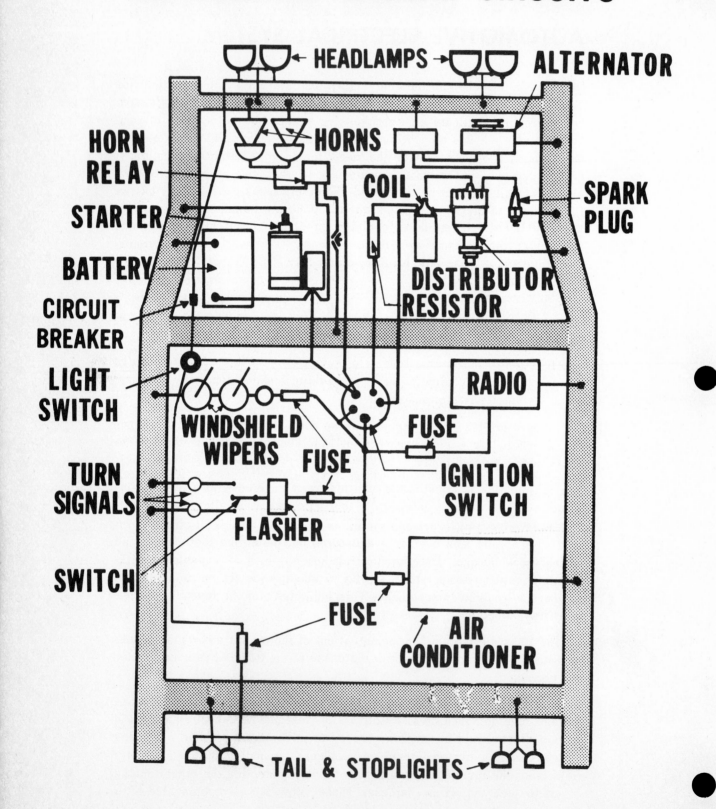

HEADLAMPS

ALTERNATOR

HORNS

HORN RELAY

STARTER

BATTERY

CIRCUIT BREAKER

LIGHT SWITCH

TURN SIGNALS

SWITCH

WINDSHIELD WIPERS

FLASHER

FUSE

FUSE

COIL

DISTRIBUTOR RESISTOR

SPARK PLUG

RADIO

FUSE

IGNITION SWITCH

FUSE

AIR CONDITIONER

TAIL & STOPLIGHTS

AUTOMOTIVE GROUND CIRCUITS

Every individual circuit in the automotive electrical system has both an insulated circuit and a ground circuit. The insulated circuit contains the source of electrical power.

The ground circuit, of equal importance to the insulated circuit, contains the ground connections and the metal parts of the vehicle that serve to return the current flowing in the circuit back to its point of origin, either the battery or the alternator or both. The metal parts of the vehicle are the frame, the firewall, the body and the ground straps.

The ground straps are an essential part of the ground circuit since the engine and body are mounted on rubber mounts or biscuits. These rubber mounts absorb engine and road vibration and noise and prevent their being transmitted into the body. The ground straps are usually braided metal straps that are bolted between the engine and firewall and between the engine and the body or fender sheet metal.

All electrical system testing is divided into two groups — testing the insulated circuit and testing the ground circuit. Proper functioning of both circuits is essential to the efficient operation of every electrical unit. Examples of these test procedures will be covered in detail as the course progresses.

WIRING DIAGRAM

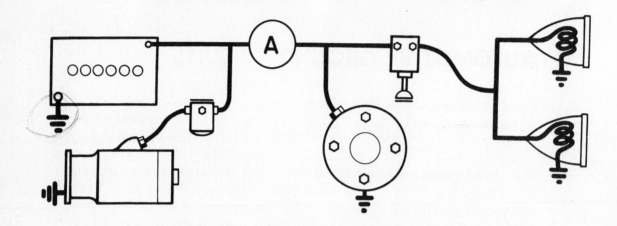

SCHEMATIC DIAGRAM

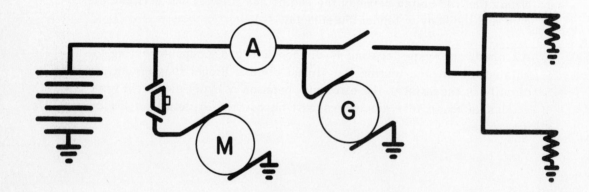

WIRING AND SCHEMATIC DIAGRAMS

In order to fully understand the illustrations that are found in the shop repair manuals, we should know the two basic types of diagrams. A wiring diagram is a pictorial view of an electrical circuit showing the components as they are actually connected.

A schematic diagram makes use of symbols to designate what the components are in the circuit. These symbols are used for clarification purposes and ease of understanding.

THE STARTING CIRCUIT

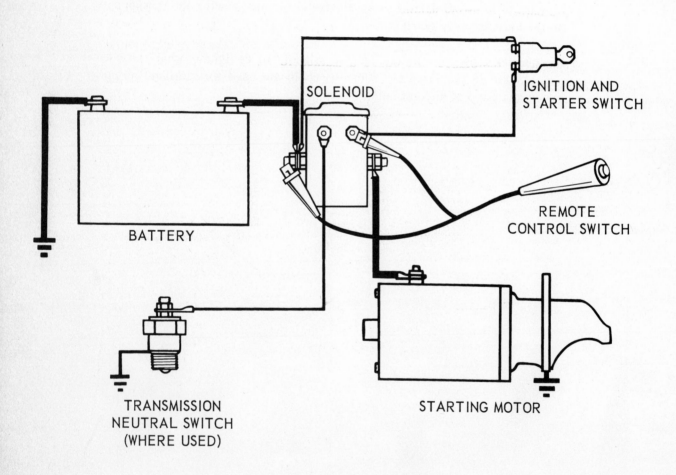

SOLENOID

IGNITION AND
STARTER SWITCH

BATTERY

REMOTE
CONTROL SWITCH

TRANSMISSION
NEUTRAL SWITCH
(WHERE USED)

STARTING MOTOR

THE STARTING CIRCUIT

A starting motor, starting switch, battery, and cables, which comprise the starting circuit, provide the power for cranking the engine. The motor receives electrical power from the battery and converts it into mechanical power which is transmitted to the engine through a drive pinion gear and the flywheel ring gear. The starting switch controls the operation by making and breaking the circuit between the battery and the motor.

The battery, starting motor, starting switch, and the wiring are all designed for the high current flow needed to produce efficient cranking power. The condition of these components is extremely critical, as even a small amount of resistance can cause a marked reduction in cranking ability. This indicates the necessity for testing the various components in the starting system.

The starter is connected in series with the battery. This is known as the high-amperage circuit. The solenoid is the switch between the battery and the starter, and is activated by a low current carrying circuit known as the control circuit.

When using a solenoid starter switch (remote control switch), **follow the hook-up instructions** that came with the switch to avoid the possibility of damaging the starter control circuit on the vehicle. For example, if the remote control switch lead is clipped to the solenoid terminal of the transmission neutral switch, the neutral switch will be burned out when the remote control switch is actuated. Also observe the precautions for remote control switch use that are covered in the Compression Test of the Tune-Up Procedure near the back of this text book.

STARTING MOTOR AMPERAGE DRAW TEST

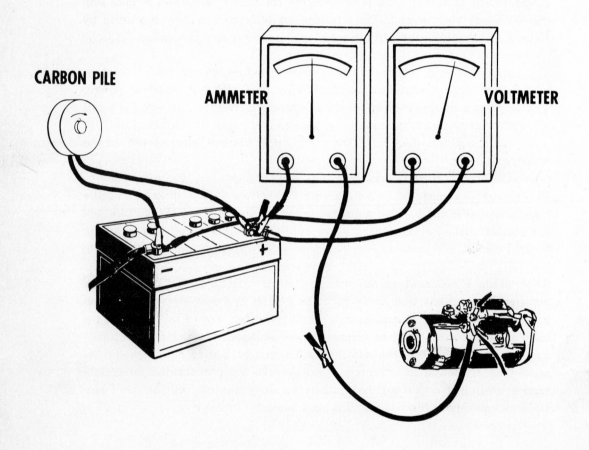

CARBON PILE

AMMETER

VOLTMETER

STARTING MOTOR AMPERAGE DRAW TEST

The starting motor amperage draw test is an on-the-vehicle check conducted with a Battery Starter Tester to detect trouble in the starting motor. If the starting motor is operating properly its amperage draw will be within specifications, 150 to 200 amperes for example, and the cranking speed will also be normal, for example 180 to 250 rpm.

So that this test can be considered a true test of the starting motor, other influencing variables will be considered to be within limits. The battery is more than ¾ charged; the cable clamps are clean and tight; the engine oil is of the recommended viscosity; and the engine temperature is normalized. Further, it is considered that the starter solenoid connections are tight, the ignition switch is functioning properly and the starter insulated and ground circuits are normal.

The Battery Starter Tester is designed with a high-reading ammeter, a voltmeter and a carbon pile which is a high-capacity variable resistor.

The starting motor amperage draw test is conducted as follows:
- Connect the tester leads.
- Connect a jumper lead from the distributor primary terminal to ground to prevent the engine from starting. Some manufacturers recommend disconnecting the primary lead from the coil, others suggest grounding the coil secondary lead.
- Crank the engine and accurately observe the EXACT voltage indicated on the voltmeter.
- Without cranking the engine, turn the tester carbon pile control until the voltmeter again reads the EXACT voltage it did while the engine was being cranked. Then read the ampere flow on the ammeter and release the tester carbon pile control. The ammeter reading is the starting motor amperage draw and should be within the manufacturer's specifications.

Higher than normal current draw, usually associated with slow cranking speed, is an indication of mechanical or electrical trouble in the starting motor. Remove the starter for repair or replacement.

Lower than normal current flow, also associated with slow cranking speed, indicates high resistance. This condition is caused by poor connections in the field or armature circuits; poor brush and commutator contact; or a defective commutator condition. Starting motor removal for service or replacement will be required.

As a word of caution—do not be too quick to condemn the starting motor unless you are certain the other factors, previously mentioned, are known to be functioning properly.

STARTER INSULATED CIRCUIT TEST

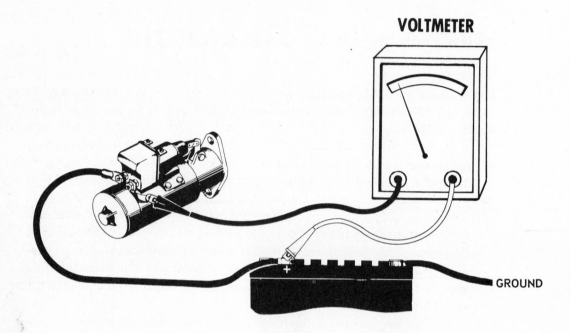

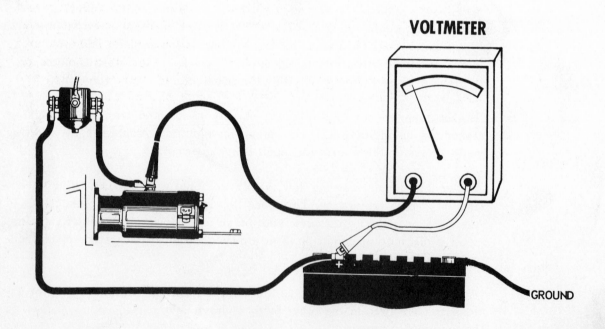

STARTER INSULATED CIRCUIT TEST

Loose or dirty connections or defective cables represent a power loss between the battery and the starter. Circuit resistance tests are made to determine if the insulated cable, switches, and ground connections can carry the current demanded by the starter. This resistance is indicated by a voltage drop. VOLTAGE DROP is the voltage expended in overcoming resistance in a given circuit. Permissible voltage drop in the average starter insulated circuit is .3 volt. The voltage drop allowed in this circuit is .1 of a volt per cable or switch. If the resistance is excessive, the result will be an extreme power loss in the starting system. When a starter motor is in operation, the high-amperage draw magnifies this seemingly low resistance value. This greatly reduces the efficiency of the entire starting system. Therefore, the circuit resistance tests are performed while the system is under normal cranking load.

To test the insulated circuit, one voltmeter lead is connected to the insulated battery terminal and the other voltmeter lead is connected to the large armature terminal on the starting motor or solenoid. Crank the engine and observe the voltmeter. If the voltage drop for the entire circuit is not excessive, objectionable resistance does not exist and no further testing is necessary. If the test results indicate excessive resistance, separate detailed tests of each component in the circuit must be conducted.

STARTER GROUND CIRCUIT TEST

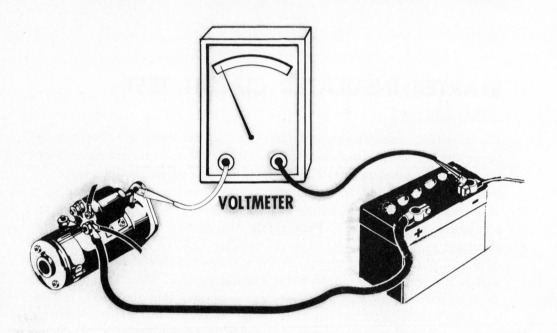

VOLTMETER

SOLENOID SWITCH
CIRCUIT RESISTANCE TEST

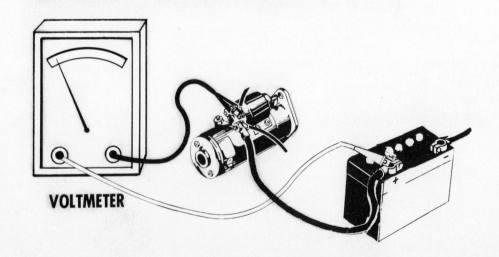

VOLTMETER

STARTER GROUND CIRCUIT TEST

To test the starter ground circuit, connect one voltmeter lead to the ground post of the battery, and the other lead to a good ground on the starter motor. Operate the starter motor and note the voltage drop. Permissible voltage drop is .2 volt.

To make the solenoid circuit resistance test, connect one voltmeter lead to the insulated battery post and the other lead to the proper control circuit terminal on the solenoid. Again note the voltage drop with starting system under load. Permissible voltage drop is .2 volt.

A higher than permissible voltage drop reading indicates at least one point of high resistance in the circuit. Separate detailed tests of each component and each connection in the circuit must be conducted. When the point of high resistance is located, the poor connection must be cleaned or tightened or the defective unit or cable must be replaced, as the case may be.

COMMONLY USED ELECTRICAL TERMS

CIRCUIT
CONDUCTOR
INSULATOR
AMPERE
VOLT
OHM
WATT

ELECTRICAL TERMS

In order to expertly diagnose electrical system troubles, the tune-up specialist must understand the electrical terms commonly used. So that a student can readily grasp the meaning of these terms, a water analogy is used, comparing the movement of electricity through a wire to the flow of water through a pipe.

A CIRCUIT is a path through which current can flow. Current flows through a circuit much like water flows through a pipe. The principle requirement of any circuit is that it must form a complete path. In tracing circuits, it is important to start at the source of electric power, either the battery or the alternator, then follow the path of current flow through the components of the insulated circuit, and return to the source through the ground circuit. A circuit is NOT complete if the current cannot return to its source.

A CONDUCTOR is a material that will pass electrical current efficiently just as a clean pipe is a good conductor for water. The ability of a conductor to carry current not only depends upon the material used but also on its length, its cross-sectional area and its temperature. A short conductor offers less resistance to current flow than a long conductor. A conductor with a large cross section will allow current to flow with less resistance than a conductor with a small cross section. For most materials, the higher the temperature of the material, the more resistance it offers to the flow of electrical current.

An INSULATOR is a material that will not pass current readily. An insulator is used to prevent leakage of electrical current.

An AMPERE is a unit of measurement for the flow of a quantity of electrical current. In terms of water analogy, this would be compared to gallons.

A VOLT is a unit of measurement of electrical pressure, or electromotive force. Voltage is sometimes described as a difference of potential between the positive and negative terminals of a battery or generator. In terms of water analogy, this pressure would be compared to pounds per square inch. In order for current to flow through a circuit, voltage must be applied to the circuit.

An OHM is a unit of electrical resistance opposing current flow. Resistance varies in different materials and varies with temperature. In terms of water analogy, this resistance would be compared to a restriction in a pipe.

A WATT is a unit of electrical power, and is obtained by multiplying volts and amperes. As a point of interest, 746 watts are equal to one mechanical horsepower.

OHM'S LAW

1 VOLT IS NECESSARY TO PUSH 1 AMPERE THROUGH 1 OHM OF RESISTANCE.

$$AMPERES = \frac{VOLTS}{OHMS}$$

$$VOLTS = AMPERES \times OHMS$$

$$OHMS = \frac{VOLTS}{AMPERES}$$

OHM'S LAW

Ohm's Law, a basic electrical rule, states that one volt (of pressure) is required to push one ampere (of current) through one ohm (of resistance).

This fundamental rule is applicable to all electrical systems and is of outstanding importance in understanding electrical circuits. It is used in circuits and parts of circuits to find the unknown quantity of voltage, current or resistance when the other two quantities are known.

Using Ohm's Law, the unknown quantity is determined as follows:

To find the amperes — divide the voltage by the resistance.

To find the voltage — multiply the amperes by the resistance.

To find the resistance — divide the voltage by the amperage.

Remember this — the current that flows in an electrical circuit is the balance between the applied voltage and the total circuit resistance.

It will not be necessary for you to stop and compute electrical values, using Ohm's Law, during a tune-up. It is advisable, however, that you have a basic understanding of its application. Your test equipment works out these problems for you, giving you the answers in the form of meter indications. With the assistance of the equipment, your attention is quickly directed to the source of the trouble.

As a general automotive electrical system trouble shooting rule, remember this - if the voltage remains constant, as it usually does except in the case of a discharged battery, an increase or decrease in current flow can only be caused by a change in resistance.

(6,24)

NEVER run battery voltage → 12 volts through an Ohm meter.

ELECTRICAL CIRCUITS

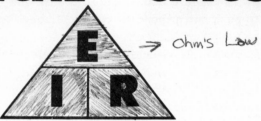

→ Ohm's Law

SERIES CIRCUIT

PARALLEL CIRCUIT

SERIES PARALLEL CIRCUIT

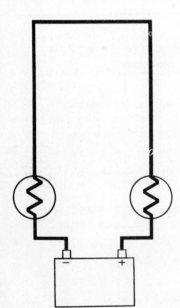

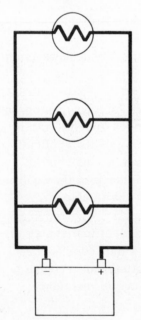

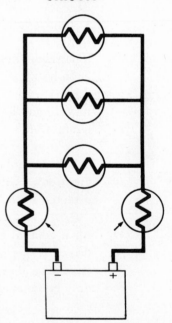

ELECTRICAL CIRCUITS

The symbol "E" represents Electro-Motive Force (Electron moving force) commonly referred to as VOLTS or electrical pressure.

The symbol "I" represents INTENSITY or current flow in AMPERES.

The symbol "R" represents RESISTANCE which is measured in OHMS. The symbol (Ω) for Omega, the last letter in the Greek alphabet, is used as the ohm symbol to avoid using the letter O which can easily be mistaken for the numeral 0 (zero).

In summation: E is volts; I is amperes; R is ohms.

The automotive electrical system is a combination of interrelated circuits. Many of the electrical components in a system have self-contained circuits. Diagnosing trouble will require a knowledge of where to look when certain conditions are indicated. The ability to trace a circuit will be of great value in pinpointing the difficulty.

SERIES CIRCUIT   A series circuit is a circuit where there is only one path in which the current can flow. Any number of lamps, resistors, or other devices having resistance can be used to form a series circuit. The total resistance of a series circuit is the sum of the individual unit resistances. The more resistances that are added to the circuit, the higher will be the total resistance. Since there is only one path for current to flow in a series circuit, this means that all current must pass through each resistance in the circuit. If an opening occurs in any portion of a series circuit, the circuit will become inoperative. This will result in an incomplete circuit. A good illustration of this circuit is the old style Christmas tree lights. If one bulb burns out, it opens the circuit and the rest of the bulbs go out.

The current flow in a series circuit is controlled by the total resistance of the circuit and the voltage applied. The current flow (amperes) will be the same in all places in the circuit. If two ammeters are connected in different places in a series circuit, both ammeters will read alike. If more resistance is added to the circuit, the amperage will become less, and if resistance is removed from the circuit, the amperage will increase.

As voltage moves current through a resistor, some of the force is expended, resulting in a loss or drop in voltage. This "voltage drop" always accompanies current moving through a resistance. Therefore, in a series circuit, the total voltage will always equal the sum of the "voltage drop" across the individual resistance units. The total voltage or the voltage across each resistance can be measured with a voltmeter, and this method called "voltage drop test" is widely used to determine circuit conditions.

PARALLEL CIRCUIT The circuit that has **more than one** path for current is called a parallel circuit. Parallel resistances connected across a voltage source have the same voltage applied to each resistance. The resistance of the individual units may or may not be the same value. Since the current divides among the various branches of the circuit, the current through each branch will vary, depending upon the resistance of the branch. However, the total current flow will always equal the sum of the current in the branches. The total resistance of a parallel circuit is always less than the smallest resistance in the circuit. If a break occurs in a parallel circuit, the circuit is not rendered inoperative because there is more than one path for current to flow back to its source. An illustration of this is street lights. If one bulb burns out, the others remain lit.

An important thing to remember in a parallel circuit is that the voltage applied remains constant at each branch.

SERIES PARALLEL CIRCUIT Many practical applications in the electrical system of the automobile depend upon a combination of series circuit and parallel circuit. This is called a SERIES PARALLEL circuit. Such combinations are frequently used, particularly in electric motors and control circuits.

MAGNETISM

MAGNETISM & PERMANENT MAGNETS

MAGNETIC FIELD

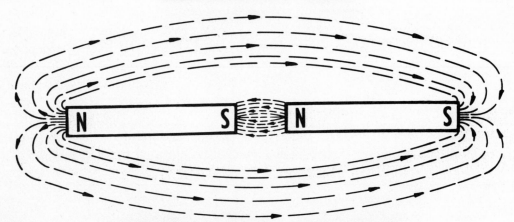

UNLIKE POLES ATTRACT

LIKE POLES REPEL

MAGNETISM

Approximately 70% of automotive electrical devices use the principle of magnetism. It is important, therefore, that you understand some of the basic laws involved.

Magnetism is an invisible force which attracts certain metals. The area that is under the influence of this magnetic force or flux is called a magnetic field. The strength of the magnetic field is governed by the number of lines of magnetic force it contains.

A magnet has a polarity known as a North Pole and a South Pole. The magnetic flux or field travels from the South Pole of the magnet internally to the North Pole and then externally from the North Pole to the South Pole. In other words, magnetic lines of force always flow out of the North Pole and into the South Pole.

The polarity of magnets becomes evident when two magnets are placed close to each other with unlike poles opposite each other. The two magnets are drawn together by the action of the combined fields making one large magnetic field.

If the magnets are held close together with their like poles opposite each other, the magnets tend to repel each other, with each magnet maintaining its own magnetic field.

You will see how magnetism is used to operate electrical units as this course proceeds.

RESIDUAL MAGNETISM

UNMAGNETIZED

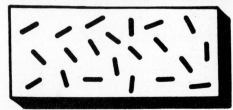

MAGNETIZED

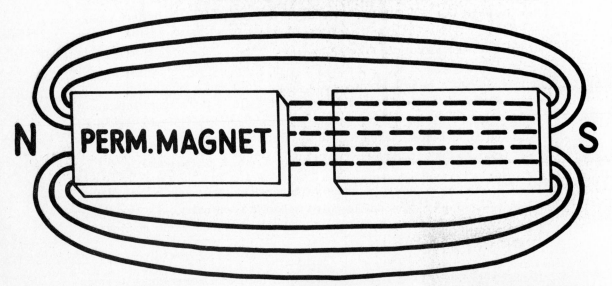

RESIDUAL MAGNETISM

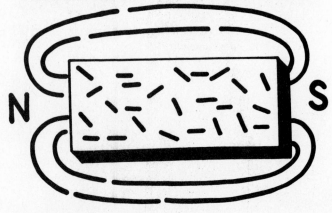

RESIDUAL MAGNETISM

Soft iron will become magnetized when placed in a field of a magnet but will lose most of its magnetism when removed from this field.

When soft iron is placed in a magnetic field and then removed, only a few of the molecules will remain in magnetic alignment. These few molecules will produce a very weak magnetic field. This is known as residual magnetism.

Residual magnetism is the factor that makes it possible for the DC generator to start its generating cycle. It is a form of self-excitation, without which, the DC generator would not function once it had been stopped.

ELECTROMAGNETIC FIELDS

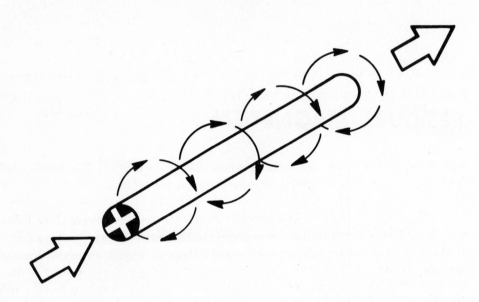

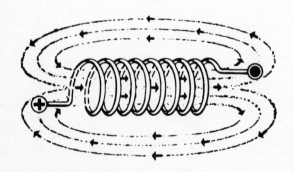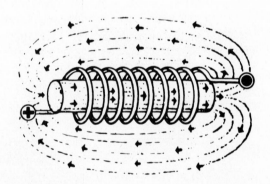

MAGNETIC FIELD SURROUNDING A CURRENT CARRYING CONDUCTOR

ELECTROMAGNETIC FIELDS

Electricity and magnetism are two separate but closely related forces. This is demonstrated by the fact that magnetic lines of force are produced around magnets, and also around conductors carrying electrical current. When electrical current is passed through a conductor, there will always be a magnetic field surrounding the conductor. The strength of this magnetic field depends upon the amount of current flow. The higher the amperage, the greater the magnetic strength.

If two conductors are arranged side by side and current passes through both conductors in the same direction, the magnetic field around each conductor will be in the same direction. As a result, the two magnetic fields will combine to form one stronger field surrounding both conductors. This causes the two conductors to be drawn together or attracted to each other. If the current is in opposite directions, the magnetic fields surrounding the two conductors will oppose each other and result in a repelling action. This is the principle involved in the operation of an electric motor such as a starter motor on a vehicle.

If a conductor is wound into a coil, the current passing through it will flow in the same direction in all turns. The magnetic field produced by each turn combines with the field produced by adjacent turns, resulting in a strong continuous field lengthwise around and through the coil. The polarity of the field produced by the coil depends upon the direction of current flow, and the direction in which the coil is wound. The strength of the magnetic field depends upon the number of wire loops and the amount of current passed through the coil This is known as AMPERE-TURNS.

The strength of the magnetic field around the coil can be materially strengthened by placing a core of soft iron inside the coil. Because the iron is a much better conductor for the magnetic lines of force than is the air, the field becomes more concentrated and much stronger. Electromagnetic relays using this basic design are used in many applications in the electrical system of the automobile.

FUEL SYSTEM

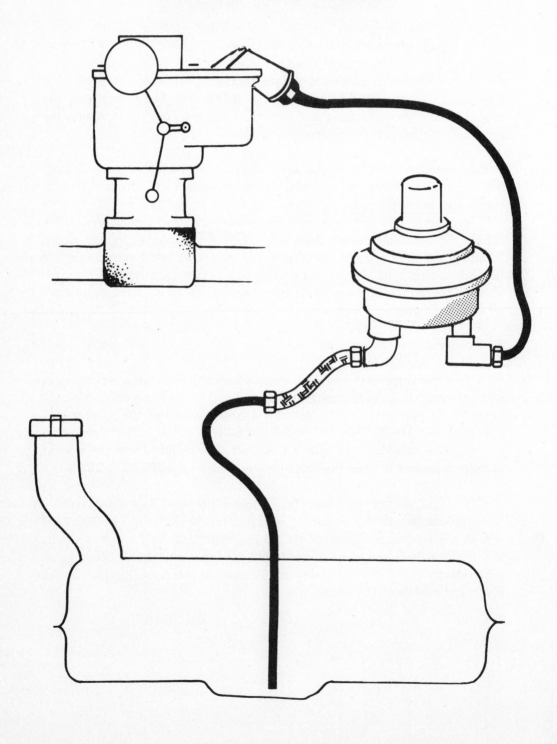

FUEL SYSTEM

The fuel system has several jobs to do. The first is to store fuel so that it is available for use when needed. Another job is to deliver this fuel to the carburetor. Still another is to mix the fuel with the proper amount of air regardless of speed or load conditions, and to control the amount of fuel and air entering the engine to meet the desires of the driver and the needs of the engine.

The components of the fuel system are fuel tank, fuel line, flex line, fuel pump, fuel filter, carburetor, air cleaner, and intake manifold.

The fuel tank is a reservoir built into the vehicle to contain a reasonable supply of gasoline for the fuel system. It must be gasoline tight so that no fuel is lost and it also must have an air vent either in the tank cap or through the Fuel Evaporation Emission Control System. Venting is extremely important since the lack of an air vent or a plugged air vent will very often cause what appears to be a fuel pump failure. Most gasoline tanks are constructed so that the gasoline pickup is a short distance above the bottom of the tank. This permits rust, water and other foreign matter to collect on the bottom of the tank but not be pulled into the gas line and through the fuel pump.

The fuel line consists of three basic parts: the pipe or tube from the gas tank to the flexible hose, the flexible hose from the gasoline pipe to the fuel pump and the pipe from fuel pump to the carburetor. The purpose of the flexible hose is to prevent damage to the gas line from engine vibration. Engines are mounted on resilient rubber engine mounts and consequently the vibration between the engine and the frame would soon cause a break in a metal gasoline line.

In operation, the fuel pipe from tank to hose and the hose from tank line to pump are subjected to vacuum or pressure below atmospheric pressure. Consequently, leaks in either of these will result in air entering the gasoline before it reaches the fuel pump. Another point to consider is that the flexible hose may collapse and partially or wholly restrict the flow of fuel from the tank to the pump. The fuel line from the pump to the carburetor is under constant pump pressure.

The fuel pump is a device which draws gasoline from the fuel tank and pumps it up into the carburetor fuel bowl. The pump must supply a sufficient quantity of fuel, at the correct pressure, to the carburetor at all times to meet the engine fuel requirements under all operating conditions.

The function of the carburetor is to mix the right amount of fuel with the right amount of air. It must deliver, at the dictates of the driver, this carefully compounded air-fuel mixture to the engine under all conditions of speed and load.

Gasoline is a mixture of a number of compounds called hydrocarbons. It is composed of about 15% hydrogen and about 85% carbon. These substances unite with the oxygen in the air at the time combustion occurs, changing the hydrogen and carbon, primarily into water, carbon monoxide and carbon dioxide. (Further discussion of this subject is covered under Exhaust Emission Control Systems.) The burning of the gasoline generates high pressures in the combustion chambers which is exerted on the heads of the pistons thereby powering the engine.

Gasoline has a potential energy three times greater than TNT and considerably greater than dynamite. Gasoline burns only when it is exposed to air. If an open can is filled with gasoline and the gasoline is lighted there will be no explosion. The surface of the gasoline will merely burn and the heat in the gasoline will be used up very slowly. In order for an explosion to occur with the gasoline, and an explosion is really a rapid burning process, more surface of the gasoline must be exposed to the air. This is why the carburetor is designed to spray or atomize the gasoline into the air stream so that the heat energy contained in the fuel can be more completely liberated for maximum power development.

FUEL PUMP

FUEL PUMP

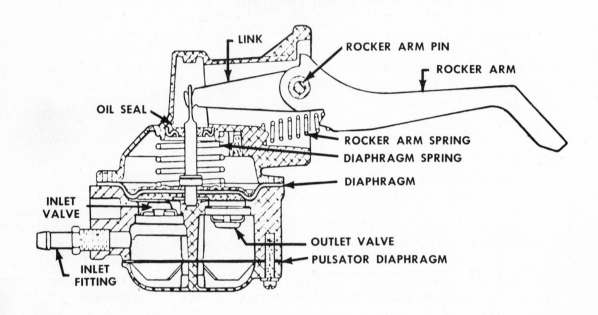

LINK
ROCKER ARM PIN
ROCKER ARM
OIL SEAL
ROCKER ARM SPRING
DIAPHRAGM SPRING
DIAPHRAGM
INLET VALVE
OUTLET VALVE
PULSATOR DIAPHRAGM
INLET FITTING

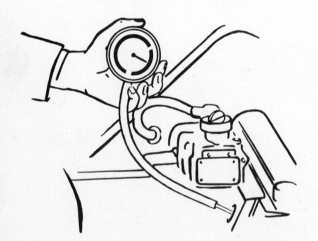

Pressure Test

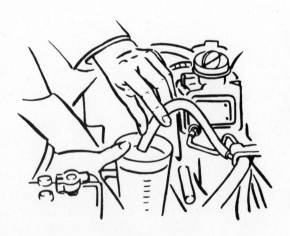

Volume Test

FUEL PUMP

The mechanical fuel pump is operated by the action of an eccentric on the camshaft which actuates the pump rocker arm. The rocker arm spring holds the arm in constant contact with the camshaft eccentric.

As the pump rocker arm is moved downward by the action of the eccentric, it bears against a shoulder on the link which pivots on the rocker arm pin. The link swings upward pulling the diaphragm upwards and compressing the diaphragm spring. The action of the rising diaphragm creates a vacuum in the fuel chamber located under the diaphragm. This action pulls the outlet valve closed and causes fuel from the gasoline tank, which is under atmospheric pressure, to enter the pump fuel chamber, which is under reduced pressure, through the inlet valve.

Further rotation of the camshaft eccentric permits the rocker arm to swing upward due to the action of the rocker arm spring. The arm then releases the diaphragm link. It cannot force the link downward because it works in an elongated slot in the link.

The compressed diaphragm spring then exerts pressure on the diaphragm which in turn applies pressure to the fuel in the chamber below. The maximum pressure exerted on the fuel depends entirely upon the strength of the diaphragm spring. The pressure on the fuel forces the inlet valve closed while forcing the fuel out of the outlet valve into the carburetor.

Pump action delivers fuel to the carburetor only when the pressure in the pump outlet line is less than the pressure exerted by the diaphragm spring. When the demand for fuel is low, and the carburetor float chamber is filled, the carburetor float presses the needle valve on its seat shutting off the entrance of more fuel from the pump.

At this time the pump builds up pressure in the fuel chamber until it overcomes the pressure of the diaphragm spring. This results in almost complete stoppage of further movement of the diaphragm until more fuel is needed.

The pulsator diaphragm in the bottom of the pump provides a form of air pocket which is compressed by the pressure on the fuel. When the pump is on the suction stroke and the pressure in the pulsator chamber is relieved, the compressed air will push the fuel out of the outlet valve as soon as the carburetor needs more fuel. The pulsator diaphragm also minimizes the pulse surges that are experienced by normal pump action.

A fuel pump is tested for pressure and volume. A typical example of a test specification is:

> Pressure: 5¼ - 6½ pounds at idle rpm and at 1000 rpm
> Volume: 1 pint in 30 seconds or less at idle rpm

While conducting fuel pump tests, OBSERVE ALL FIRE PRECAUTIONS.

FLOAT SYSTEM

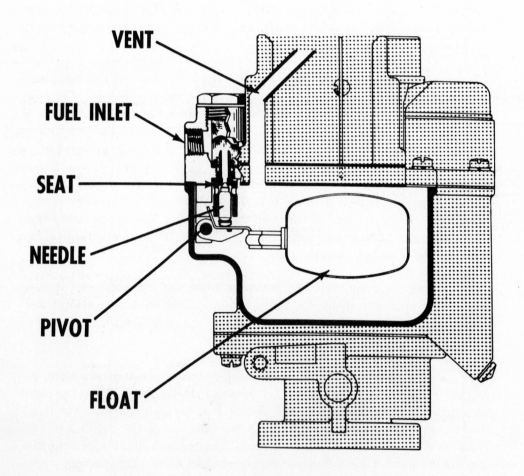

VENT

FUEL INLET

SEAT

NEEDLE

PIVOT

FLOAT

FLOAT SYSTEM

The automobile is driven at variable speeds under a variety of operating conditions. Each condition has its own specific fuel requirement. The carburetor is designed with several different systems or circuits, each with its own particular function, to supply these varying needs. These systems or circuits are: the float system; idle circuit; main metering system; power system; accelerating system; and the choke system. Each system will be discussed in turn.

The float system maintains a constant supply of fuel in the float chamber for immediate use by the fuel metering systems in the carburetor. Fuel, under pressure from the fuel pump, enters the float chamber through the fuel inlet line. This fuel raises the float on its pivot thereby controlling the needle valve and admitting only enough fuel to replace that being used. As the fuel level drops due to the fuel being consumed by engine operation, the float lowers permitting fuel pump pressure to unseat the needle valve and admit more fuel to the bowl. The entrance of fuel again raises the float. When the correct level is reached, the float again closes the needle valve on its seat shutting off the fuel supply from the pump. Actually the needle valve will drop only slightly as the fuel is consumed since the action of the valve is quite sensitive.

The level at which the float maintains the fuel supply in the float chamber is extremely important. If the correct fuel level is 3/16 inch below the tip of the main discharge nozzle, an increase of 1/32 inch would decrease the distance that the fuel has to be lifted by 16 percent, resulting in a fuel mixture that is too rich. A decrease of 1/32 inch in fuel level would increase the distance that the fuel has to be lifted by 16 percent, resulting in a fuel mixture that is too lean.

The float chamber is vented to the atmosphere through a port in the air horn. The function of the vent is to allow the fuel to be smoothly withdrawn from the chamber into the various carburetor circuits.

An Idle Circuit is needed that will deliver some gasoline and air when the throttle is closed

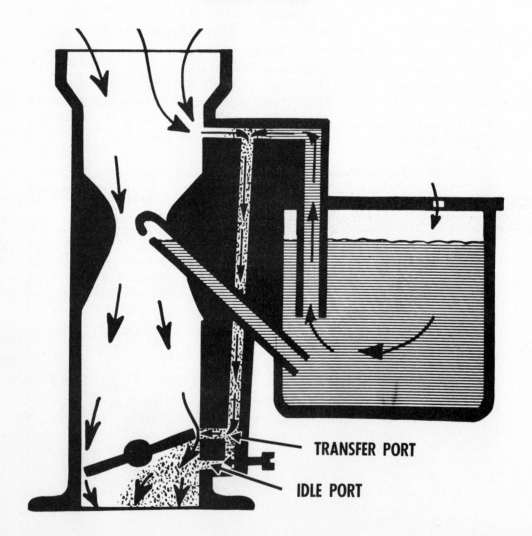

TRANSFER PORT

IDLE PORT

IDLE CIRCUIT

When the throttle valve is closed there is not enough air flow through the carburetor to create a low-pressure area at the main discharge nozzle. Therefore, an idling system has been provided. The idling system delivers fuel through the idle discharge port located below the throttle valve. A drop in pressure is created at this point because the throttle valve acts as a pressure dropping device when in the closed position. All the air that passes through the carburetor throat must pass around the throttle valve. The fuel is forced through this idle system because of the pressure difference between the atmospheric pressure on the fuel in the fuel bowl and the low-pressure area created by the nearly closed position of the throttle valve.

Fuel drawn from the float bowl is metered by an idle jet and mixed with a controlled quantity of air from one or more calibrated air bleeds. The air-fuel mixture is then discharged from the idle port just below the slightly open throttle mixing with air passing through the carburetor to form the idle mixture. An adjusting needle at the idle port can be turned to provide a leaner or richer air-fuel ratio.

As the throttle is opened, additional air flows through the carburetor. Although still insufficient to draw fuel from the main nozzle, this increased air flow will result in an excessively lean mixture and will cause a flat spot in engine performance unless additional fuel is made available. This problem is overcome by adding a transfer port just above the closed throttle position. When the throttle is opened it gradually exposes the transfer port to intake manifold vacuum causing discharge of air-fuel mixture from this port as well as from the idle port.

Both the idle and transfer ports are carefully designed to provide a smooth transition between idling and cruising speeds of 20 to 40 miles per hour depending upon carburetor design. At these speeds, the throttle opening and air flow are sufficient to permit the main metering system to supply the necessary fuel. The idle and low-speed system now cease to function due to insufficient vacuum acting on the idle system ports.

Complete carburetor adjustment is covered in the Tune-Up Procedure Section of this course.

MAIN METERING SYSTEM

VENT

VENTURI

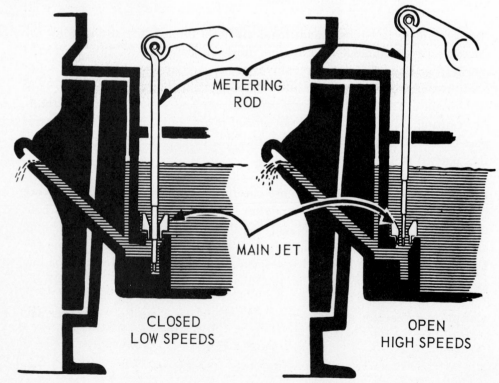

METERING ROD

MAIN JET

CLOSED
LOW SPEEDS

OPEN
HIGH SPEEDS

MAIN METERING SYSTEM

Carburetors are designed with a venturi tube positioned in the main carburetor body. The function of this tube is to control the fuel discharge from the main jet by restricting the air opening through the carburetor.

The air taken into the carburetor air horn must pass through the venturi and in passing through this restricted area the air momentarily increases in speed. This increase in air speed causes a partial vacuum or low-pressure area at the narrow point of the venturi. This low-pressure area is utilized to cause the fuel to flow because of the differential pressure between atmospheric in the fuel bowl and lower than atmospheric in the venturi. As the throttle is opened, the speed of the air flow through the carburetor is increased with a proportional pressure drop in the venturi.

Since maximum velocity and pressure drop are obtained at the smallest cross section of the venturi, this is the logical point to locate a discharge nozzle for maximum fuel flow. Air entering the carburetor passes through the venturi where its velocity is increased and its pressure is reduced. Atmospheric pressure in the bowl forces fuel through the nozzle into the air stream. To prevent fuel flow from the nozzle when the engine is not operating, the fuel level in the bowl is maintained slightly below the discharge nozzle opening.

The amount of fuel metered into the air stream is controlled by a main jet or metering orifice. The metering orifice is calibrated in size to provide accurate metering of the fuel in the economy range. In most cases this metering orifice is not large enough to carry a sufficient amount of fuel to supply the engine at high speeds and full load. The metering rod type of system combines both the economy and power ranges. To accomplish this a jet sufficient in size to supply the required amount of fuel under full load operation is used. To reduce the effective size of the orifice for part throttle operation, a calibrated rod called a metering rod is inserted in the jet. This rod is mechanically connected to the throttle shaft and is raised as the throttle is opened.

To provide the proper opening in the orifice at any given throttle position, the metering rod is graduated in various diameters along its lower end, that portion which moves in the orifice. At idle or low-speed operation, the metering rod is inserted in the jet so that the large diameter of the rod is in a position to permit only a minimum of fuel flow through the orifice. As the throttle is opened, the metering rod is raised in the jet, progressively increasing the effective orifice opening, thereby, allowing a greater flow of fuel. At wide open throttle, the metering rod is raised to its limit in the orifice, allowing maximum fuel flow for high-speed and full-load operation.

VACUUM CONTROLLED METERING ROD

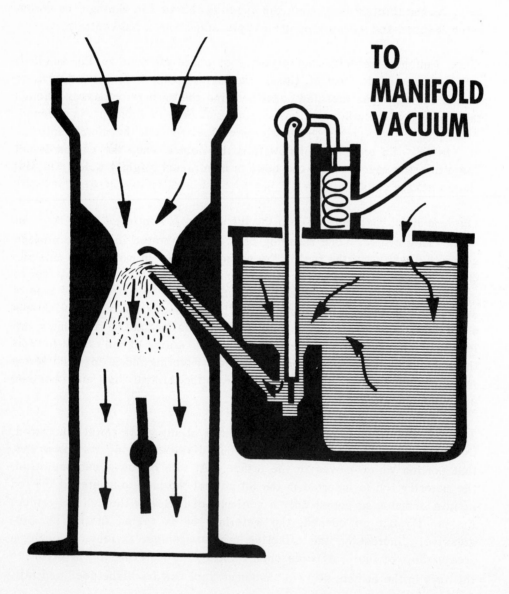

TO MANIFOLD VACUUM

VACUUM CONTROLLED METERING ROD

The vacuum controlled metering rod provides more metering rod movement with less throttle action than the manually controlled type metering rod.

When the throttle is closed or nearly closed, engine manifold vacuum is high. Atmospheric pressure pushes the piston down, fully compressing the spring. The metering rod provides maximum restriction in the main jet and very little gasoline flows through the high-speed nozzle.

As the throttle is opened slightly, there is a corresponding decrease in manifold vacuum. The spring pressure then raises the piston and metering rod slightly permitting an increased fuel flow out of the high-speed nozzle.

When the throttle is opened wide, manifold vacuum drops sharply. The spring pressure pushes the piston and metering rod upward to its full limit of travel. This permits a maximum flow of gasoline through the high-speed nozzle to supply the demand for full power.

During closed or part throttle operation, the high engine manifold vacuum keeps a larger portion of the metering rod inserted in the main jet as previously explained. On quick throttle opening the sudden drop in manifold vacuum permits the spring to push the piston and metering rod upward very quickly. This action permits increased fuel flow to supply the sudden demand for increased power.

Every performance demand made on the engine by the action of the throttle is immediately compensated for by the sensitive action of the vacuum controlled metering rod.

POWER JET OPERATION

TO MANIFOLD VACUUM

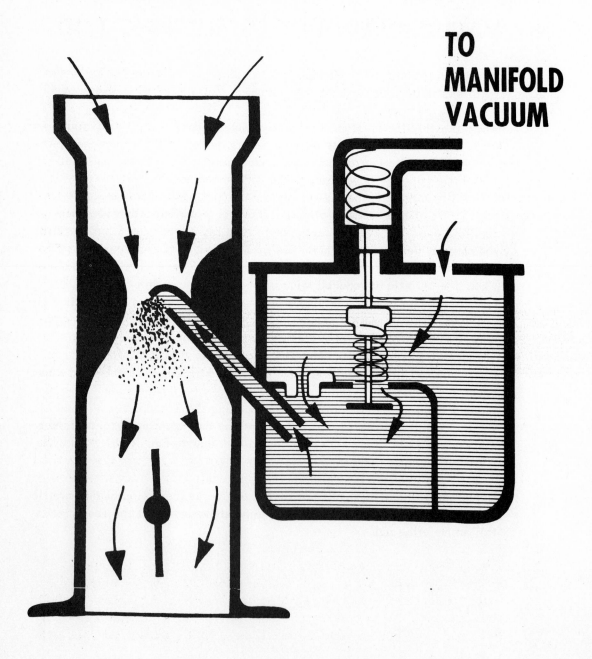

POWER SYSTEM

For normal cruising speeds an economical mixture of about sixteen parts of air to one part of fuel are calibrated by the main metering system. For high-speed or full throttle operation a richer mixture is necessary. This mixture must be approximately twelve parts of air to one part of fuel. The power system automatically supplies the added fuel when it is required by the engine.

The power system may be said to serve two functions. During normal cruising speeds the power system does not operate and may be called the economizer system. Under the demands of full throttle operation the system acts as a fuel enriching device or power system to satisfy the demands for increased power.

The power system has two important parts; a piston which is a part of the air system, and a valve which is a part of the fuel system. The power system is actuated by the manifold vacuum which reflects a true indication of the power requirements of the engine. The vacuum is taken from a point located below the throttle plate and is transmitted through a passage to the top of the power piston. The bottom of the piston is exposed to atmospheric pressure of the air in the bowl.

During idle and low-speed operation, manifold vacuum creates a low-pressure area above the piston causing the piston to be forced upward by the atmospheric pressure in the fuel bowl. This action closes the power jet and compresses the calibrated spring above the piston. The power system is not operating at this time.

When high speed or load requirements are placed on the engine, a greater opening of the throttle plate is necessary to maintain speed. When the throttle plate is opened, less resistance is offered to the engine suction and the manifold vacuum is reduced. The drop in vacuum permits the piston spring to push the piston down thereby opening the power jet and permitting the entrance of the additional fuel necessary to meet the engine's demands.

The power valve is located in the bottom of the float chamber and is covered by fuel at all times. Fuel from the float chamber flows through the power valve which acts as a supplementary main jet. The amount of fuel allowed to flow through the power system is metered by restrictions. This restriction prevents the power system from delivering too much fuel at high speeds.

THE ACCELERATING SYSTEM
puts a charge of gasoline into this air space until the gasoline starts from the nozzle.

ACCELERATING SYSTEM

When the throttle is suddenly opened a volume of uncharged air tends to enter the engine. This condition is caused by the fact that the fuel is heavier than the air and is consequently harder to quickly set into motion. Besides, the fuel has a tendency to cling to the wall of the main well resisting a sudden increase in movement. A lack of fuel exists therefore, when transferring quickly from the idle system to the main system.

To enable the engine to accelerate rapidly without hesitation, an accelerating pump temporarily supplies the extra fuel necessary for more power until the main metering system can catch up with the needs of the engine. If it were not for the accelerating pump, a large volume of uncharged air would enter the engine and the engine would "flat spot" or slow down momentarily.

The accelerating pump is linked to the throttle lever. The pump is so constructed that when the throttle is moved to the closed position, the piston in the pump chamber moves upward drawing a quantity of fuel from the fuel bowl past a check valve and into the pump chamber. When the throttle is opened, the piston moves downward forcing the fuel out of the pump chamber into the carburetor venturi where it mixes with the uncharged air. This enriched mixture discharge covers the time lapse required for the main metering system to come into full operation.

The discharge of the fuel occurs instantly when the throttle is opened. However, a slot in the piston stem allows an overriding action of the linkage. This arrangement subjects the piston to spring tension which provides a prolonged discharge of the fuel rather than just a single spurt.

The accelerating system uses a simple pump employing check valves to permit fuel to be taken in and discharged in accordance with the movement of the pump piston. As the piston moves upward in the cylinder, the inlet check ball is lifted off its seat allowing fuel to be drawn into the chamber from the fuel bowl. The discharge valve seats itself during the chamber loading operation thus preventing air from the venturi being drawn in through the discharge nozzle.

As the throttle is opened, the pump piston is forced downward and the pressure on the fuel causes the inlet check ball to seat itself thus preventing the fuel from returning to the fuel bowl. The fuel also moves through the discharge passage lifting the pump discharge valve off its seat and allowing the fuel to move out of the discharge nozzle into the carburetor throat.

CHOKE ACTION

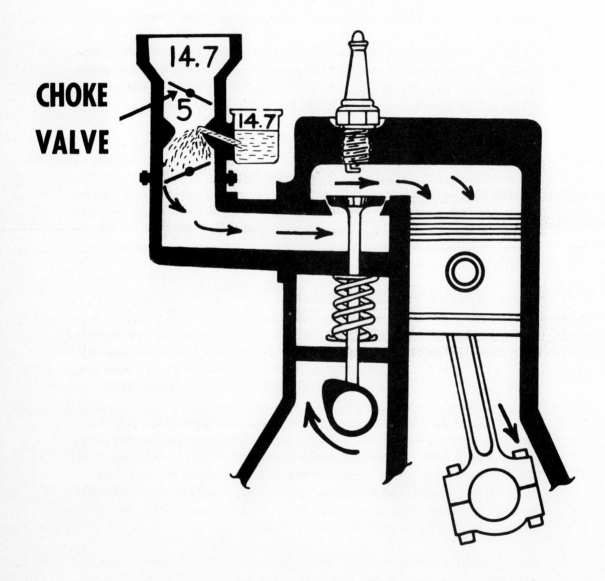

CHOKE VALVE

14.7

5

14.7

14.7

CHOKE

When the engine is cold, a richer mixture than the idle system can supply, is required for starting. A choke valve is used to accomplish this. With the choke valve closed, atmospheric pressure is present above the choke valve. As the engine is turned over by the starter, a low-pressure area is created just below the choke valve. As an example, the chart shows a pressure of only five pounds (equal to 10 inches of vacuum) below the choke valve. This permits air pressure of 14.7 pounds per square inch to push gasoline out of the main nozzle even though very little air is flowing past this main nozzle. This provides a rich mixture for starting the engine.

A rich mixture is required for starting because with a cold engine atomization of the gasoline is not very efficient and much of the gasoline condenses on the walls of the intake manifold. Consequently, all of the gasoline coming out of the main nozzle is not carried into the cylinders with the air.

The choke valve is mounted on a shaft which is off center. This is to provide a self-opening action of the choke as the engine starts and more air rushes in to meet the needs of the engine. Most chokes are spring loaded or held closed by spring tension. When the engine starts, the choke valve can open sufficiently so that the mixture is not overrich and the engine will continue to run. With the spring loaded choke valve, if the throttle valve is closed and less air flows into the engine, the spring tends to close the choke valve again maintaining about the same pressure differential above and below the choke valve even though the air volume past the choke varies with engine speed.

AUTOMATIC CHOKE
PISTON TYPE

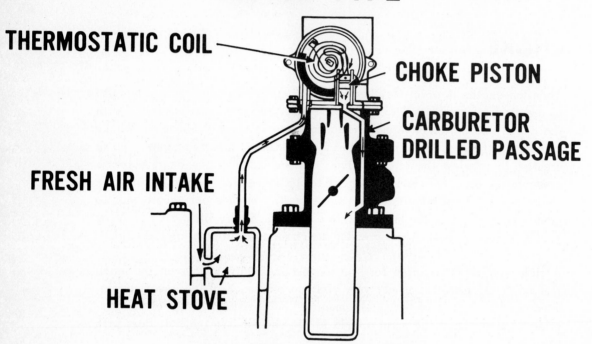

THERMOSTATIC COIL

CHOKE PISTON

CARBURETOR DRILLED PASSAGE

FRESH AIR INTAKE

HEAT STOVE

DIAPHRAGM TYPE

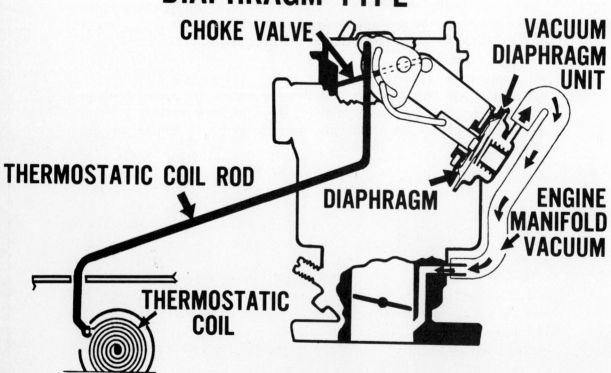

CHOKE VALVE

VACUUM DIAPHRAGM UNIT

THERMOSTATIC COIL ROD

DIAPHRAGM

ENGINE MANIFOLD VACUUM

THERMOSTATIC COIL

AUTOMATIC CHOKE

PISTON TYPE

The automatic choke is controlled by a combination of a thermostatic bi-metal coil spring and a vacuum operated piston. Both the spring and piston are linked to the choke valve. During cold starting the coil of the spring holds the choke closed. As soon as the engine starts, intake manifold vacuum acting on the piston opposes the spring action thereby tending to open the choke. As the engine warms up, heated fresh air drawn from the exhaust manifold through a tube causes the thermostatic spring to relax, permitting the choke valve to fully open. Failure of the choke to open properly will cause an excessively rich mixture, fouling of the spark plugs and poor engine performance.

To counter the opening action of a choke by air velocity a thermostatic spring tends to hold the choke valve closed. This spring is usually constructed of a strip of brass and a strip of steel fused together and wound into a coil. Since these materials have a different coefficient of expansion, the spring will exert a greater tension when cold than when hot. When the engine is cold, the spring exerts a force on the choke valve to hold it in a closed position. As the engine warms up, this tension gradually decreases in proportion to the increase in engine temperature. Therefore, when the engine is warm, spring tension is relaxed permitting the choke valve to open.

The vacuum piston is connected by linkage to the choke shaft and operates in a cylinder which is connected by a passage to a point below the throttle valve. This makes the piston subject on one side to the low pressure of the intake manifold. When the choke is moved to the closed position by the thermostatic spring, the piston is moved to one end of the cylinder in which it operates. Vacuum or low pressure is applied to the piston the instant the engine starts. The piston moves toward the low pressure and partially opens the choke. As the piston moves in the cylinder, it encounters slots cut in the side of the cylinder. The piston will only move toward the low-pressure area to a point where the slots provide a passage around the piston. The low pressure is then satisfied by air moving from the atmospheric pressure at the heat stove end of the pickup, through the choke housing, which includes the thermostatic spring, and then around the piston by way of the passage formed by the slots in the piston cylinder. The result is that the choke is allowed to assume a "starting" position as determined by the thermostatic spring, and at the instant the engine starts the vacuum operated piston provides a "running" position of the choke by opening the choke to a definite minimum position. The piston linkage being connected to the choke shaft results in the vacuum piston cancelling part of the tension of the thermostatic spring.

As the thermostatic spring continues to uncoil, it simply moves the piston down in its cylinder. Vacuum during this period is bypassed through the cylinder slots.

A choke lock, which is part of the fast idle and choke linkage, is used to keep the choke in the off position should the throttle be held open for an extended time. This will prevent the choke from closing due to the loss of warm air flow through the choke housing caused by the drop in engine vacuum.

A device, called the choke unloader, is connected to the throttle lever and choke linkage. This mechanism permits the choke to be opened sufficiently, by flooring the accelerator, to clear the excess gasoline from the engine should the engine flood during starting.

Automatic chokes of this type are adjusted by tensioning the bimetal spring. This is effected by loosening the cover retaining screws and turning the cover to align the emboss mark on the cover with the proper index mark on the choke body and then tightening the retaining screws. Specifications for an automatic choke setting are, for example: index mark, one notch rich or two notches lean.

As previously stated, air from outside the engine is heated as it passes through the heat stove on its way to the automatic choke. In the event the hot exhaust gases burn out the heat stove, exhaust gases will flow into the automatic choke presently resulting in carbonizing of the choke passages and mechanism. When this conditon is detected, the choke mechanism must be cleaned and adjusted and the heat stove replaced.

DIAPHRAGM TYPE (VACUUM BREAK)
Some late model carburetors as the Rochester BV, 2GV and 4MV, the Stromberg WWC3, the Carter AVS and YF models, and Holley 2209, 4150 and 4160 models use a diaphragm type automatic choke. In this choke design, the vacuum diaphragm replaces the choke piston and the thermostatic coil spring is mounted in the intake or exhaust manifold, over the heat crossover passage.

Engine vacuum below the throttle plate is used, in conjunction with the thermostatic spring, to operate the diaphragm choke in much the same manner as in the piston type choke.

The adjustment of the diaphragm type choke is performed by disconnecting the choke rod, at its upper end, from the carburetor choke lever. Hold the choke valve closed and lift the choke rod upward until it strikes its stop in the thermostat housing. Then check the position of the top of the rod in relation to its hole in the choke lever. The average setting is from ½ to 1 rod diameter above the top of the lever hole. If an adjustment is necessary, bend the choke rod, as required, in its offset area to affect the desired adjustment. Connect the choke rod to the choke lever and check for free choke operation and for complete choke closing.

A choke setting equal to "2 notches" rich in the piston type choke would be basically 2 rod end diameters above the hole in the choke rod.

AUTOMATIC CHOKE

Some automatic chokes use the heat of the engine coolant instead of the heat of the exhaust gases to operate the choke. The engine coolant circulates through hoses to the choke water housing where its heat is transmitted to the bi-metal thermostatic spring thus influencing the choke valve position.

ELECTRIC CHOKE ASSIST SYSTEMS

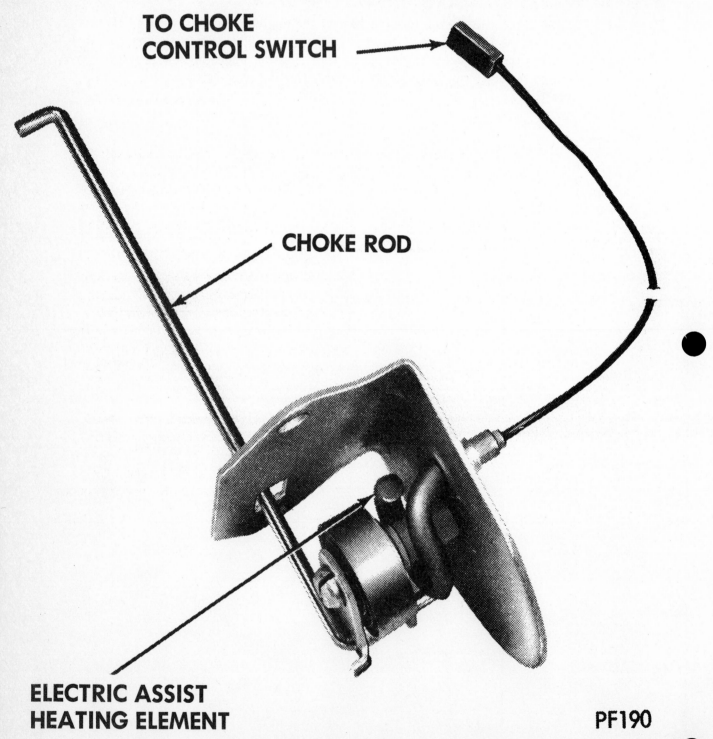

TO CHOKE CONTROL SWITCH

CHOKE ROD

ELECTRIC ASSIST HEATING ELEMENT

PF190

The heating element of an electric assist choke assembly.

ELECTRIC CHOKE ASSIST SYSTEMS

To hasten the warmup process and shorten the period during which the engine is on fast idle, modern engines use an electrically heated choke mechanism. This is an electrical heating element contained within the thermostatic choke housing to supply additional heat to the thermostatic choke spring.

Most of these systems have a thermally operated switch that cuts off the heater current when the ambient temperature is below a certain level, for instance, 60°F. This allows normal choke action, and a longer fast idle period, during cold weather operation. In warmer weather, the cutoff switch closes to allow the heating coil to shorten the choke operating period.

Troubleshooting these systems is usually done with an ohmmeter and voltmeter. The voltmeter, by connecting it to the input terminal on the choke housing and ground, shows the presence or absence of operating voltage. In some cars, this voltage is derived from the alternator and therefore the engine must be running for this check. Check the manufacturer's test procedure to see if this voltage is controlled by ambient temperature.

The heating element itself is generally checked by measuring its resistance with a low range ohmmeter. The specified resistance varies, but typically is in the range of 4 to 12 ohms. Again, it is necessary to check the specified operating temperature of the choke cutoff switch. If the ambient temperature is below the operating point, the switch may be open, preventing a resistance measurement of the heating coil.

HOT IDLE COMPENSATOR VALVE

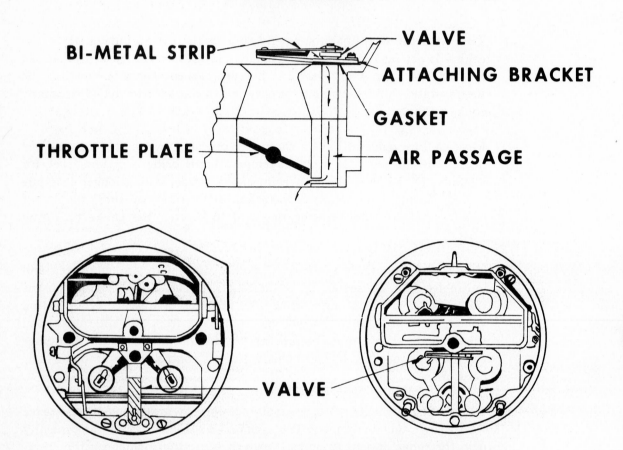

BI-METAL STRIP

VALVE

ATTACHING BRACKET

GASKET

THROTTLE PLATE

AIR PASSAGE

VALVE

INTERNAL CARBURETOR MOUNTINGS

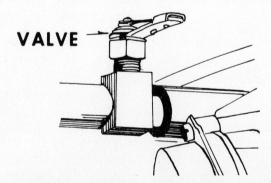

VALVE

EXTERNAL MOUNTING IN PCV SYSTEM HOSE

HOT IDLE COMPENSATOR VALVE

During periods of high underhood temperatures, as in start-and-stop driving in congested traffic on a hot day, there is a tendency for fuel vapors to collect in the intake manifold. These vapors are the result of fuel "spill-over" caused by the pressure generated by the heat on the fuel in the carburetor bowl. The result is an overly rich fuel mixture which prompts rough idle and a tendency to stall. Engines that stall under this condition are frequently very difficult to restart. The function of the carburetor hot idle compensator valve is to prevent engine stalling under these conditions by opening an auxiliary air bleed circuit. This action permits additional air to enter the manifold under the throttle plates and dilute the rich fuel mixture in the manifold.

The hot idle compensator valve is a simple mechanism mounted internally in the carburetor air horn on most 4-barrel and many 2-barrel carburetors. Some 2-barrel carburetors have the hot idle valve externally mounted in the PCV hose near the carburetor. The valve, positioned over its port, is mounted on and controlled by, a flat thermostatic spring. During periods of normal operating temperatures the valve remains closed and there is no action in the idle compensating circuit. When the air temperature in the carburetor air horn reaches approximately 120 degrees, the thermostatic spring opens the valve permitting additional air to bypass the throttle plates and dilute the rich fuel mixtures collecting in the intake manifold. When operating temperatures return to normal, the compensating valve automatically seats itself and the air bleed circuit is shut off.

The compensating valve in most carburetors is internally mounted in a horizontal position in the carburetor throat. In other carburetors, Ford and Autolite particularly, the valve is mounted on its side, so to speak. In any event, the valve is quite obvious when its size, shape and function is understood.

The hot idle compensator valve is nonadjustable and no attempt should ever be made to bend the thermostatic spring as a method of adjustment. Valve replacement is the only recommended procedure in the event of valve malfunction.

In most carburetor adjustment procedures there are several qualifying requirements among which is the statement: "Hot idle compensator valve held closed." Since the carburetor adjustment is one of the last settings made during a tune-up and the engine may have been idling for some time, the hot idle compensator valve may be open.

If the carburetor is adjusted with the valve open, the mixture adjustment will be made to compensate for this additional air supply. When the engine temperature returns to normal and the valve closes, the idle mixture will be too rich and the idle will be rough. To prevent this condition from occurring, the valve (if it is open) should be held closed with a gentle pressure on the valve using a pencil or a finger tip, or similarly blocking off its air passage, while the carburetor is being adjusted.

MANIFOLD
HEAT CONTROL VALVE

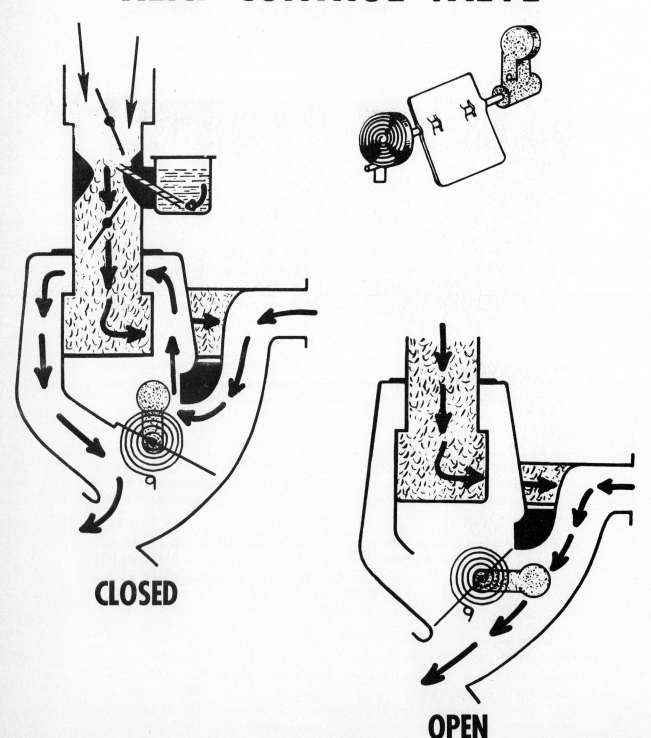

CLOSED

OPEN

MANIFOLD HEAT CONTROL VALVE

The intake manifold depends on heat from the exhaust manifold to assist in proper fuel vaporization. A thermostatically-controlled valve arrangement provides the correct heat transfer between the exhaust manifold and the intake manifold. The thermostatic spring in the automatic choke is actuated by heat from the exhaust manifold. If intake manifold temperatures do not rise in proportion to the exhaust manifold temperatures, an unbalanced condition between the automatic choke and the intake manifold temperature will result. For instance, if the intake manifold is retarded in reaching operating temperature due to improper heat control valve operation, the engine will have the effect of running lean because of insufficient vaporization of the fuel. Meanwhile, the exhaust has caused the automatic choke to open providing a still leaner mixture.

When the engine exhaust manifold is cold, the thermostatic spring holds the heat control valve in a closed or "heat on" position. In this position the exhaust gas is deflected upward and around the intake manifold hot spot, then downward into the exhaust pipe. Once the intake manifold has reached its normal operating temperature, the heat control valve is moved to the open or "heat off" position by the action of the counterweight and the thermostatic spring which has been heated by the hot exhaust gases. When the valve is in the open position, the exhaust gas is deflected directly from the exhaust manifold into the exhaust pipe.

If the manifold heat control valve sticks in the open position, engine warm-up is delayed because the desired preheating action of the fuel charge does not take place. This means the choke stays in operation for a longer time increasing fuel consumption and diluting the motor oil.

If the heat valve sticks in the closed position, the fuel charge is constantly subjected to preheating even after engine temperature is normalized. This condition results in a lean fuel mixture, loss of power, hard "hot engine" starts, and possible burning of the engine valves.

The valve should be checked for free operation at every tune-up. It should move freely approximately 90 degrees without evidence of sticking or binding. If there is a tendency of the valve to stick, it should be hand-operated while the recommended lubricant (not motor oil) is liberally applied to the valve shaft ends. If, due to obstructions, the lubricant cannot be applied directly to the valve shaft ends, it is often possible to touch a long screw driver to each shaft end and let the lubricant run down the screw driver shank.

If the valve cannot be moved by hand, lightly tap the shaft ends with a small hammer to break the shaft loose. Then apply the lubricant while hand-operating the valve. A valve that will not respond to service must be replaced.

INTRODUCTION TO VEHICLE EMISSION CONTROL SYSTEMS

The automobile engine is considered one of the major contributors to the pollution of the atmosphere. It has been established that the partially burned hydrocarbons, carbon monoxide and nitrogen oxides contained in the automotive exhaust, when in the presence of sunlight, creates "photochemical smog." When the smog concentration is sufficiently high, it has definite eye and lung irritating effects. Smog is a coined word - taken from the words smoke and fog.

To combat this serious condition Federal legislation has been passed that requires every passenger car and light truck engine, beginning with the 1968 models, to be equipped with Federally-approved emission control systems. These systems have been used, by virtue of state law, in California since 1966.

There are basically three sources of emissions from the automobile engine:

1. Crankcase vapors. These vapors are composed of certain amounts of compressed fuel charge and exhaust fumes that pass the piston rings as "blow-by" and collect in the crankcase. These vapors represent approximately 20% of the engine's emissions.

2. Exhaust emissions. These emissions contain: 1. Hydrocarbons which are basically unburned fuel; 2. Carbon monoxide which is an invisible, odorless poisonous gas caused by overly rich fuel mixtures as during periods of choke operation; 3. Oxides of Nitrogen. These compounds are a product of the oxygen and nitrogen in the atmosphere formed under the high temperatures and pressures in the engine's combustion chambers. These three by-products make up about 60% of the engine's emissions.

3. Fuel evaporative emissions. Evaporation of the fuel from the carburetor and fuel tank take place constantly and contribute about 20% of the vehicle's emissions. The release of these fumes is highest during periods of "heat soak" immediately after engine shut-down. Devices and systems to control this condition have been introduced on the 1970 vehicles sold in California and have been installed nationally on all cars starting with the 1971 models (since the enactment of Federal control regulations).

CRANKCASE EMISSION CONTROL

The emissions of crankcase vapors have been effectively controlled by the Closed PCV System. This system is standard on all engines, nationwide, including Imported cars, beginning with the 1968 models and has been standard equipment on all California engines since 1964.

EXHAUST EMISSION CONTROL

The emissions from the engine's exhaust system must, by Federal law, be held within specified limits. These national standards apply to all vehicles sold in the United States beginning with the 1968 models. The standards apply equally to Imported car models sold in the U.S. The Federal stand-

88

ards established in 1968 set limits on hydrocarbon and carbon monoxide emissions. These standards have been getting progressively tighter and it is projected they will become even more stringent by 1975 by which time oxides of nitrogen and particulate emission standards will also have been established.

The Federal Government has approved several different car manufacturers' systems designed to control exhaust emissions. For purpose of identification these systems may be basically classified into two catagories: engine modification and air injection.

The engine modification systems employ a degree of engine redesigning that permit a greater reduction of objectionable emissions within the engine's cylinders by promoting more complete combustion. In addition, these systems also employ various assist units to still further effect emission reductions.

The air injection systems employ a belt-driven, positive displacement, low-pressure air pump that continuously pumps air through a system of hoses and tubes into the cylinder head area of each engine exhaust valve. When the exhaust valve opens the injected fresh air supplies the additional oxygen to support further combustion of the unburned hydrocarbons in the exhausting gases. In this manner the hydrocarbons are consumed in the engine instead of being exhausted into the atmosphere. The air injection system also incorporates various assist units to aid in controlling objectionable emissions.

Since the automobile's air-polluting emissions are highest during periods of idling and closed-throttle deceleration, most emission control devices are designed to affect these two operational areas.

FUEL EVAPORATION CONTROL SYSTEMS

In addition to systems designed to control crankcase emissions and exhaust emissions, the State of California established standards to control air-polluting gasoline evaporation emissions that occur from the automobile fuel tank and carburetor. This standard applied to all 1970 cars sold in California and has been applied Federally to all cars beginning with the 1971 models.

Two evaporation control systems are used. They are essentially similar, the basic difference being the method used to store the fuel vapors. Some 1970-71 models and all 1972 models use the popular system of storing the vapors in a charcoal granule filled canister which is mounted in the engine compartment. The other system used on some 1970-71 models, stored the vapors in the engine crankcase. With either system, as soon as the engine is started the vapors are drawn from the canister, or crankcase, into the engine cylinders where they are burned.

The only way the fuel system vapors can escape is through the Evaporation Emission Control System. Air pollution from this source is thereby eliminated.

Following are detailed descriptions of each of the systems discussed in this introduction.

POSITIVE CRANKCASE VENTILATION SYSTEMS

DRAFT TUBE   **OPEN PCV SYSTEM**   **CLOSED PCV SYSTEM**

SEALED CAP

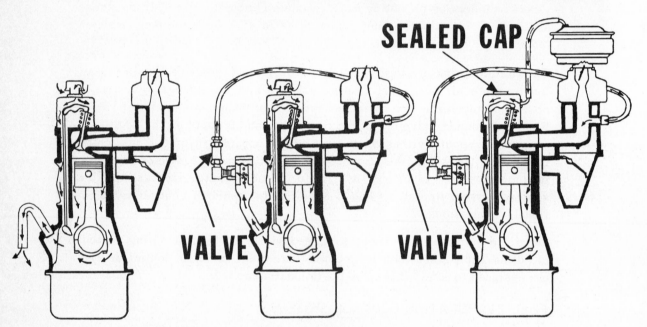

VALVE   VALVE

PCV METERING VALVE

TO INTAKE MANIFOLD

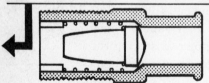 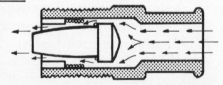 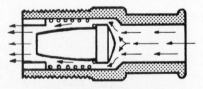

ENGINE STOPPED (VALVE CLOSED)   IDLING OR LOW-SPEED AIR FLOW RESTRICTED (VACUUM HIGH)   HIGH SPEED MAXIMUM AIR FLOW (VACUUM LOW)

POSITIVE CRANKCASE VENTILATING SYSTEMS

During engine operation a certain amount of the fuel charge and exhaust gases find their way past the piston rings into the crankcase. These gases are commonly called "blow-by."

The presence of blow-by gases and moisture condensation in the crankcase causes contamination of the motor oil. This combination of elements results in a sludge formation in the oil. In a neglected state, this sludge can block the oil pump screen and oil galleries with disasterous results. Further, these contaminants are highly corrosive and will eventually result in acid etching and rusting of highly polished internal engine surfaces. Varnish and lacquer formation on critical engine parts, as hydraulic lifters and cam lobes, will seriously interfere with efficient engine operation. Proper ventilation of the crankcase to remove these fumes, gases and condensed vapors is essential to prevent motor oil contamination and extend engine life.

DRAFT TUBE
Ventilating the crankcase by passing air through it has been the standard practice for many years. Air entered the crankcase through the oil fill cap, or crankcase breather as it was frequently called, passed through the crankcase picking up the fumes and gases, and exhausted from the engine through a road draft tube.

This system had two disadvantages. First, the amount of air passing through the crankcase is governed by how fast the car is being driven. The movement of air under the vehicle is the factor that created a greater or lesser pressure differential at the end of the draft tube with the air pressure at the breather cap. At idle and low cruising speeds the passage of air through the crankcase is inadequate for effective ventilation. Second, the fumes ventilated from the crankcase are exhausted in to the atmosphere thereby contributing to air pollution.

OPEN POSITIVE CRANKCASE VENTILATING SYSTEM
The Positive Crankcase Ventilating System, frequently referred to as "PCV", assists in reducing the air pollution caused by automobile crankcase vapors. In the PCV system the draft tube is eliminated. Blow-by gases and vapors are drawn from the crankcase by intake manifold vacuum, through tubing into the intake manifold and into the engine cylinders where they are burned. The rate of air flow through the system is controlled by a valve or "Jiggle pin" as they are sometimes called. A few systems employ a fixed metered orifice instead of a valve. A slight valve movement meters the air flow to provide the proper degree of crankcase ventilation consistent with engine demands and vehicle speeds.

With the elimination of the major portion of the blow-by gases, engine oil contamination is reduced, water vapor, rust and corrosive elements are largely eliminated and engine life is materially extended. All these benefits are secured while air pollution is considerably reduced.

When service of the PCV system is neglected, the valve usually ceases to function since it is in the line carrying the crankcase contaminants. When this happens the system is blocked, crankcase pressure developes, and the crankcase fumes and blow-by gases are forced out of the crankcase backwards through the breather cap. Not only are the benefits of the PCV system lost but the crankcase fumes are again polluting the atmosphere. This undesirable factor led to the development of the "Closed" PCV System.

CLOSED PCV SYSTEM

In the Closed PCV System the crankcase breather cap is replaced with a solid type cap similar to a gas tank cap. Air entering the system is introduced through the carburetor air cleaner. Air flow through the system is controlled by the action of a valve influenced by manifold vacuum the same as in the standard PCV System. However, should the closed system be neglected to the extent that crankcase pressure causes a reverse flow of blow-by gases and fumes, they will be drawn into the carburetor through the air cleaner. The smog-producing fumes will not be vented into the atmosphere but will be burned in the engine.

The Closed PCV System was made mandatory by Federal law and has been standard equipment on all passenger car and light truck engines starting with the 1968 models.

Testing and servicing procedures for the PCV System will be covered in the Tune-Up Procedure Section of this course.

EXHAUST EMISSION CONTROL SYSTEMS

ENGINE MODIFICATION TYPES

EXHAUST EMISSION CONTROL SYSTEMS

ENGINE MODIFICATION TYPES

CCS-CONTROLLED COMBUSTION SYSTEM
IMCO-IMPROVED COMBUSTION SYSTEM
ENGINE-MOD-ENGINE MODIFICATION SYSTEM

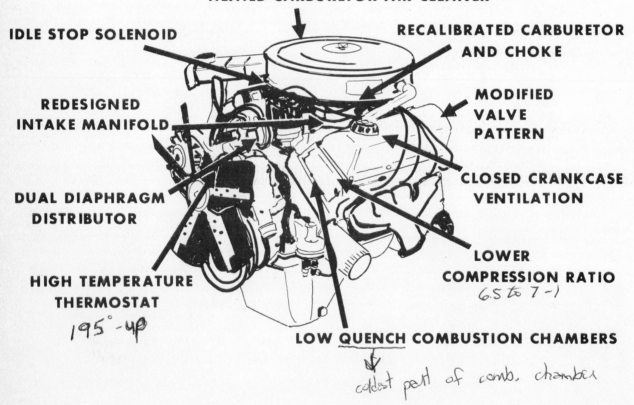

HEATED CARBURETOR AIR CLEANER

IDLE STOP SOLENOID

RECALIBRATED CARBURETOR AND CHOKE

REDESIGNED INTAKE MANIFOLD

MODIFIED VALVE PATTERN

DUAL DIAPHRAGM DISTRIBUTOR

CLOSED CRANKCASE VENTILATION

HIGH TEMPERATURE THERMOSTAT
195°-4θ

LOWER COMPRESSION RATIO
6.5 to 7-1

LOW QUENCH COMBUSTION CHAMBERS

coldest part of comb. chamber

EXHAUST EMISSION CONTROL SYSTEMS

ENGINE MODIFICATION TYPES

Several exhaust emission systems come under the classification of Engine Modification. Each car manufacturer employing a system of this type applies his own trade name. General Motors call their system the Controlled Combustion System (CCS); Ford Motor Company use the Improved Combustion (IMCO) System; and American Motors call their system the Engine-Mod (Engine Modification) System. The Chrysler Corporation Cleaner Air System (CAS), also an engine modification type, is covered separately.

CEC

CAP - clean air package (chrysler)

Engines employing the modification type emission control system have been redesigned in several areas to more completely burn the fuel charge in the combustion chambers thereby reducing objectionable exhaust emissions that are a product of incomplete combustion. Basically, this engine redesigning includes the following features. The carburetor is specially calibrated for lean fuel mixtures. The choke is more sensitive to variations in engine temperature during engine warm-up permitting faster choke release. Higher idle speeds reduce idle emissions. Redesigned intake manifold permits smoother fuel charge movement with less restricted fuel induction while further providing better fuel distribution.

The distributor has been recalibrated for greater spark timing control. Dual diaphragm vacuum units provide spark advance in the cruising ranges and spark retard during idle and periods of deceleration. Ignition advance control systems (covered under Assist Units) are employed to prevent vacuum spark advance during acceleration through the gears and permit vacuum advance only at cruising speeds in high gear to more effectively control emissions.

A redesigned cam shaft provides a modified valve timing pattern with a greater degree of overlap in most instances. The combustion chamber contour has been modified to reduce the quench area and the compression ratios have been reduced. In some engines the compression has been reduced sufficiently to permit the use of nonleaded gasoline. Higher temperature thermostats in the cooling system are also being used.

6.5 to 7-1

In addition to the engine design features listed above these engines are also fitted with various assist units to still further control the engine's emissions.

The net result of all these improvements has been more complete combustion with cleaner running engines.

servicing cooling systems:

block-check → chemical to test for exhaust gas to see if cracks in head or block.

coolant tester

pressure tester

(Skogi's Radiator Shop)

EXHAUST EMISSION CONTROL SYSTEMS

CLEANER AIR SYSTEM (CAS)

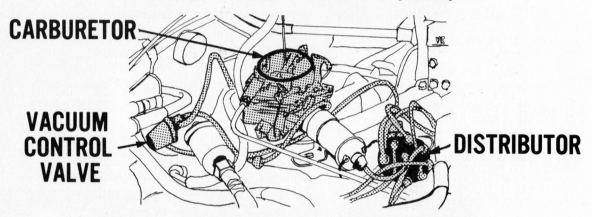

CARBURETOR

VACUUM CONTROL VALVE

DISTRIBUTOR

THREE ELEMENTS OF CAS SYSTEM

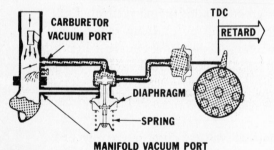

CARBURETOR VACUUM PORT

TDC

RETARD

DIAPHRAGM

SPRING

MANIFOLD VACUUM PORT

ENGINE IDLING

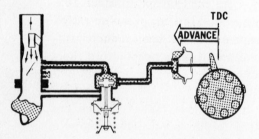

TDC

ADVANCE

ACCELERATING AND CRUISING

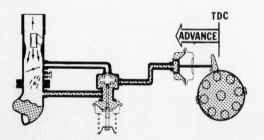

TDC

ADVANCE

DECELERATING

EXHAUST EMISSION CONTROL SYSTEMS

CLEANER AIR SYSTEM (CAS)

The Cleaner Air System, referred to as "CAS" is the Exhaust Emission Control System developed by the Chrysler Corporation. All passenger cars and light trucks built by Chrysler for sale in California in 1966-67 and nationally starting with the 1968 models are equipped with this system. On the 1966-67 models the name of the system was known as Cleaner Air Package (CAP). In the following copy, references to CAS also apply to CAP.

The Chrysler Corporation contends that a properly tuned engine has sufficiently low exhaust emissions during normal cruising and accelerating ranges. The excessive emissions occur during low speed operation and particularly during deceleration. The controlling factors designed into the CAS system are directed toward these driving ranges. The CAS equipped engine employs essentially three elements that vary from the conventional engine. They are a specially calibrated carburetor, a modified distributor, and a vacuum control valve used in conjunction with the distributor.

The specially calibrated carburetor has several unique features. The choke opens more quickly during the warm-up cycle to lean the air/fuel mixture as soon as the engine can effectively operate on the leaner mixture. The idle mixture is set to a leaner ratio and is combined with a slightly increased throttle opening at idle speed. The main jets are set to operate close to the lean setting to provide a lean mixture that burns more efficiently at cruising speeds.

The modifications to the distributor provide a greater range of breaker plate travel to acquire a substantial spark retard at idle speed while retaining the maximum range of spark advance for cruising speeds. CAS equipped engines are generally timed close to Top Dead Center.

The vacuum control unit is a vacuum-sensitive device that is connected by hoses to the carburetor, to the intake manifold, and to the vacuum advance unit on the distributor. The function of the control valve is to permit maximum distributor timing advance during periods of deceleration thereby effectively reducing exhaust hydrocarbon emissions.

Both carburetor and intake manifold vacuum act on the vacuum control valve at all times. From these two vacuum sources the control valve senses the engine speed and load conditions. It relays this message to the distributor vacuum advance unit to vary the spark timing as required. By this system of sensitive vacuum controls the exhaust emissions are reduced to an acceptable level.

During idle speeds, the control valve remains seated. The distributor vacuum advance unit, now being exposed to only a weak carburetor vacuum signal, retards the initial timing. This retard timing may be as much as 15 degrees from the conventional setting.

During acceleration and normal cruising speeds the increased carburetor vacuum causes the distributor vacuum advance unit to advance the spark timing in the conventional manner. Intake manifold vacuum still is not strong enough to overcome the control valve spring so the valve remains seated.

During deceleration, the period of highest exhaust emission, carburetor vacuum is weak due to a closed throttle, but intake manifold vacuum is very strong. The high manifold vacuum unseats the control valve and full manifold vacuum is exerted on the distributor vacuum advance unit pulling it into a position of maximum spark advance. As deceleration progresses and engine speed falls to idle, intake manifold vacuum drops, the control valve spring seats the valve, and the distributor vacuum advance unit returns to a retard position. In this manner the exhaust emissions are held to an acceptable level.

Due to engine and transmission modifications, the CAS system on most late model Chrysler-built engines does not employ the vacuum control valve.

Starting with the 1970 models, Chrysler introduced Carburetor Heated Air as part of their Cleaner Air System.

Chrysler's recommended procedure for testing the CAS system is covered in the Tune-Up Procedure Section of this course.

EXHAUST EMISSION CONTROL SYSTEMS
AIR INJECTION TYPE

EXHAUST EMISSION CONTROL SYSTEMS

AIR INJECTION TYPE

AIR INJECTION REACTOR, THERMACTOR, AIR GUARD, AIR INJECTION SYSTEM

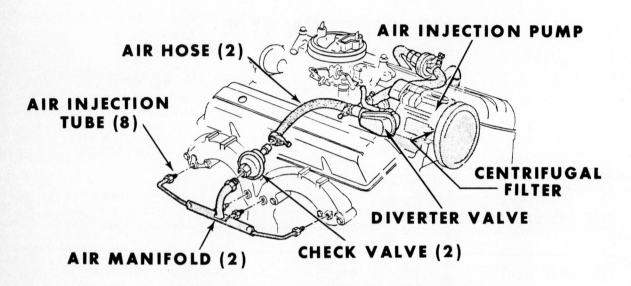

AIR INJECTION PUMP

AIR HOSE (2)

AIR INJECTION TUBE (8)

CENTRIFUGAL FILTER

DIVERTER VALVE

AIR MANIFOLD (2)

CHECK VALVE (2)

TYPICAL V-8 ENGINE INSTALLATION

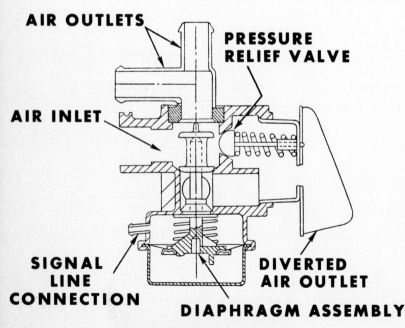

AIR OUTLETS

PRESSURE RELIEF VALVE

AIR INLET

SIGNAL LINE CONNECTION

DIVERTED AIR OUTLET

DIAPHRAGM ASSEMBLY

DIVERTER VALVE

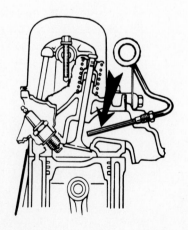

AIR INJECTION TUBE

EXHAUST EMISSION CONTROL SYSTEMS

THE AIR INJECTION TYPE

One of the Federally-approved exhaust emission control systems is the air injection type. It employs a belt-driven, positive displacement, low-pressure air pump that continuously pumps filtered air through a system of hoses, manifolds and injection tubes into the area of each engine exhaust valve in the cylinder heads. When the exhaust valve opens the injected fresh air supports further combustion and ignites the hot unburned portion of the exhausting gases. This action produces a more complete burning of the exhaust gases before they are expelled from the tailpipe. This system also uses carburetors calibrated for lean fuel mixtures and recalibrated distributors among other assist units.

Although all air injection systems have many parts in common, each car manufacturer using the system calls it by his own trade name. General Motors call the system Air Injection Reactor (A.I.R.); Ford Motor Company calls the system Thermactor (Thermal Reactor); American Motors (including "Jeep") name their system Air Guard; and Chrysler Corporation calls their system the Air Injection System.

Three types of valves are used in the Air Injection System. A relief valve, check valve(s), and a diverter valve, or an anti-backfire valve which was used on the earlier air injection systems. An air by-pass valve on some systems serves the same function as the diverter valve.

The relief valve is mounted in the diverter valve or on the air pump. Its function is to release excessive air flow and pressure directly from the pump at high speed. This air pressure release will prevent excessively high exhaust temperatures at high cruising speeds.

The check valve is positioned at the entrance of the air manifold(s). On some engines, specifically Cadillac and Buick, the air passages are cast in the cylinder heads. This design reduces the amount of external plumbing and the possibility of air leaks. The function of the check valve is to prevent the reverse flow of air and gases from the engine into the pump should engine pressures at any time exceed pump pressure. This condition would occur if the drive belt should fail, if the tailpipe became restricted or if a backfire occured.

The diverter valve has the function of preventing a "backfire" in the exhaust system during periods of sudden deceleration. When the throttle is closed suddenly at high speeds, the carburetor air is shut off quickly but due to the inertia of fluids, the gasoline continues to flow momentarily. This condition results in an excessively rich fuel mixture which might explode in the exhaust ports and manifold if air was injected. The high intake manifold vacuum that is present on sudden throttle closing lifts the diverter-valve diaphragm exposing a by-pass port through which the pump's air supply is dumped into the atmosphere instead of being pumped into the engine's exhaust ports. The muffler is designed to silence the sound of the escaping air during the periods of deceleration and high speed operation.

The anti-backfire valve, previously used, served somewhat the same function as the diverter valve. The high intake manifold vacuum that accompanies rapid deceleration triggers the anti-backfire valve so that it opens for about one second to permit a "gulp" of clean air from the pump to enter the intake manifold. This form of momentary air bleed serves to "lean out" rich fuel mixture thereby greatly reducing the backfire probability. Since the introduction of the air "gulp" also prevents a rapid drop in engine speed on deceleration, the air injection-equipped engine does require a dashpot to prevent engine stalling.

It can be readily seen that the action of the air pump is much the same as fanning the dying embers of a fire into a flame by the addition of air (oxygen) to complete the fuel's combustion. In the case of the automobile engine, the burning of the unconsumed hydrocarbons and carbon monoxide gases are rekindled in the exhaust port and exhaust manifold by the introduction of the air. This is a form of afterburn. The smog producing elements are thereby consumed in the engine instead of being released into the atmosphere.

In conjunction with the air pump, most engines also are equipped with a carburetor that is calibrated for leaner mixtures. Distributor initial timing and advance curve characteristic changes are also frequently employed.

To effectively control the crankcase emissions in conjunction with the exhaust emissions, the air injection-equipped engine is also equipped with a Closed Positive Crankcase Ventilating System. Remember, each of these systems is entirely separate from the other and each performs a different function.

Proper and periodic tuning of an engine equipped with an exhaust emission control system is particularly critical. Fouled spark plugs, improper ignition timing, incorrect carburetor adjustment and other defects can produce a greater proportion of unburned hydrocarbons than the exhaust emission system can control. The elements of properly tuning and air injection-equipped engine will be covered later in this course. The procedures for testing this system are covered in the Tune-Up Section of this Text Book.

IMPORTED CAR EMISSION CONTROL SYSTEMS
Dual Intake Manifold System

IMPORTED CAR EMISSION CONTROL SYSTEMS

DUAL INTAKE MANIFOLD SYSTEM

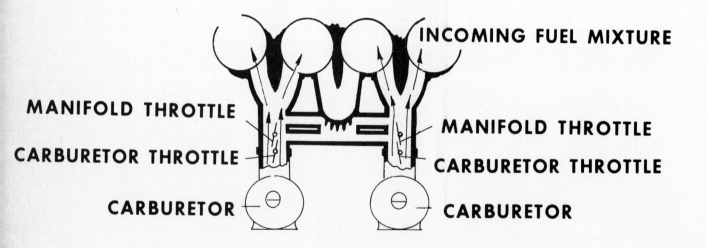

HOT EXHAUST GAS

INCOMING FUEL MIXTURE

MANIFOLD THROTTLE

MANIFOLD THROTTLE

CARBURETOR THROTTLE

CARBURETOR THROTTLE

CARBURETOR

CARBURETOR

PREHEATING CHAMBER

INCOMING FUEL MIXTURE

MANIFOLD THROTTLE

MANIFOLD THROTTLE

CARBURETOR THROTTLE

CARBURETOR THROTTLE

CARBURETOR

CARBURETOR

IMPORTED CAR EMISSION CONTROL SYSTEMS

DUAL INTAKE MANIFOLD SYSTEM
Starting with their 1968 models, the Imported car builders had to incorporate approved crankcase and exhaust emission control systems on their engines to meet Federal Emission Standards on all cars imported into the United States.

In many respects the exhaust emission systems used on Imports have much in common with our domestic systems. The use of specially calibrated carburetors and distributors and the employment of retarded spark timing and increased idle rpm are evident on several imported car models. Renault, Peugeot, Simca, Triumph, Fiat, Volkswagen and some Datsun models employ these features. Air injection systems, used in the States for four years, are employed on models of BMC cars (MG, Austin and Austin Healey), Ford, Mercedes, Porsche, Opel, Toyota and other Datsun models.

The one exhaust emission system that differs from any domestic arrangement is the Dual Intake Manifold System used on Volvo and Jaguar cars. This engine modification system uses the dual manifold principle, spark retard at idle, modified carburetion for lean fuel mixtures and increased idle speeds.

It has been definitely established that burning lean carburetor mixtures is a major factor in reducing objectionable exhaust emissions. The primary difficulty of using lean mixtures is the associated loss of acceptable engine performance. During periods of steady cruising, lean mixtures work well enough. But changes in engine load and engine speed, cold engine starts and warm-up operation, along with changes in pressure, humidity and temperature introduce variables that are difficult to control with a lean mixture. The old method of providing a richer fuel mixture to cover all these variables had to be dispensed with because of the objectionable emissions rich mixtures create. The dual manifold system is said to have been designed to provide acceptable performance through all speed ranges with lean fuel mixtures.

The dual intake manifold principle employs a centrally positioned heating and mixing chambers crossflow pipes which permit each carburetor to feed two cylinders while further acting as an intake pulsation damper and carburetor balancing device; and two throttles per carburetor, one being the primary or conventional throttle and the other being a secondary or manifold throttle.

During periods of idling and low-speed, in-town driving the manifold throttles are closed and the fuel mixture from the carburetors is directed through the heating chamber. Here the fuel charge is vaporized into a highly combustible mixture and is then passed into the engine cylinders. This preheating and vaporizing of the fuel charge is the heart of this system. It achieves a reduction in objectionable emissions while providing acceptable engine performance with the use of lean fuel mixtures.

As engine speed is increased to approximately half power output, a cam device on each carburetor throttle gradually opens the manifold throttles. A portion of the fuel charge now bypasses the heating chamber. During periods of greater carburetor throttle opening for maximum acceleration or periods of high-speed driving, the manifold throttles are completely opened and the fuel charge is inducted directly into the cylinders, completely bypassing the heating chamber.

One of the major features attributed to the dual manifold design is that the crossflow pipes automatically balance the carburetors. This is a very important factor in dual carburetor installations. It is claimed the engine could operate acceptably, except for loss of maximum performance, with only one carburetor functioning.

ASSIST UNITS

In conjunction with the different exhaust emission control systems, various externally-mounted, accessory-type units, generally referred to as "assist" units, are used to help increase driveability and promote more acceptable performance. In some instances these devices help the emission control system function with greater efficiency. Not all assist units are employed with every emission control system. But each will be covered so that its function, operation and test procedure will be understood.

Two of these units are the Heated Carburetor Air Systems and the Carburetor Idle Stop Solenoids. These two assist units are covered on the immediate following pages.

Several other assist units are closely associated with the ignition system and are covered in that section of this Text Book. They include: dual diaphragm distributors; double-acting vacuum spark control units; thermostatic vacuum switch; deceleration vacuum advance valve; distributor retard and advance solenoids; along with various vacuum spark control systems designed to limit oxide of nitrogen and hydrocarbon exhaust emissions.

Carburetor mixture limiter caps and the deceleration valve are also considered assist units and are covered in the carburetor adjustment section of this text book.

HEATED CARBURETOR AIR SYSTEMS

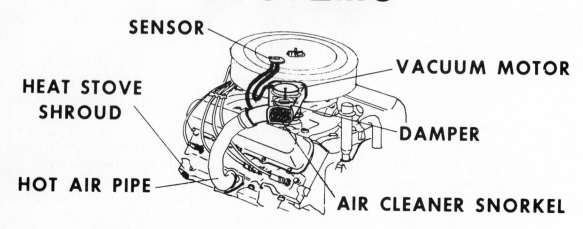

SENSOR

VACUUM MOTOR

HEAT STOVE
SHROUD

DAMPER

HOT AIR PIPE

AIR CLEANER SNORKEL

GENERAL MOTORS AND CHRYSLER TYPE

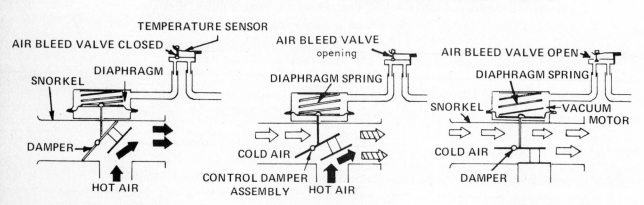

TEMPERATURE SENSOR

AIR BLEED VALVE CLOSED

DIAPHRAGM

SNORKEL

DAMPER

HOT AIR

AIR BLEED VALVE
opening

DIAPHRAGM SPRING

COLD AIR

CONTROL DAMPER
ASSEMBLY

HOT AIR

AIR BLEED VALVE OPEN

DIAPHRAGM SPRING

SNORKEL

VACUUM
MOTOR

COLD AIR

DAMPER

HOT AIR DELIVERY

REGULATING MODE

COLD AIR DELIVERY

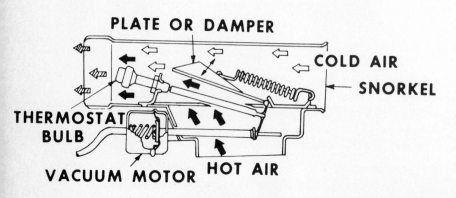

PLATE OR DAMPER

COLD AIR

SNORKEL

THERMOSTAT
BULB

VACUUM MOTOR

HOT AIR

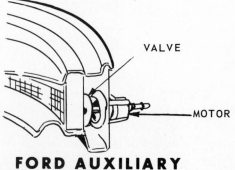

VALVE

MOTOR

FORD AUXILIARY AIR INLET VALVE

FORD AND AMERICAN MOTORS TYPE

HEATED CARBURETOR AIR SYSTEMS

The Heated Carburetor Air System (also called "Hot and Cold" Air Intake System or Climatic Combustion Control) has been introduced as another means of assisting in the control of objectionable exhaust emissions.

The function of the system is to direct heated air to the carburetor when underhood temperature is below 100° F. and up until the temperature reaches approximately 130° F. This system provides desirable emission control throughout the vehicle's operating range while further effecting increased fuel economy and improved engine warm-up.

Heating the carburetor air during periods of cold engine operation permits proper fuel vaporization with more complete combustion of the fuel mixture. It further provides for a shorter period of choke action which also has the benefit of improving engine warm-up and lowering exhaust emissions. Carburetor icing, a troublesome stalling condition prompted by the build-up of ice crystals on the throttle plate, is virtually eliminated by the intake of heated air.

A temperature sensing device in the carburetor air cleaner housing or snorkel, senses the incoming air temperature and through suitable linkage operates a valve plate or damper door. When underhood temperature is below 100° F., the valve plate is held closed (heat-on position) so that only heated air, drawn through a duct connected to a heat stove positioned around the exhaust manifold, can enter the carburetor. As engine and underhood temperatures rise, the sensing device opens the valve plate slightly permitting a mixture of heated and unheated air (at about 100° F.) to enter the carburetor. As operating temperature rises to normal (130° F. or higher) the sensing device opens the valve plate completely (heat-off position) so that only engine compartment unheated air enters the induction system.

The temperature sensing device can be a **vacuum control** unit which governs the vacuum motor action of controlling the damper door by applying or bleeding-off vacuum to the motor. General Motors and Chrysler vehicles employ this type. Or the device can be a thermostat capsule or bulb which **mechanically** operates the damper door. This is the type used on Ford and American Motors engines.

To insure acceptable engine performance during periods of cold engine acceleration when the supply of warm air may not be adequate, the damper must be momentarily fully opened to unheated air to provide an air supply that is sufficient to the engine's demands. In the General Motors and Chrysler vacuum-operated unit, the drop in intake manifold vacuum on sudden acceleration permits the vacuum motor diaphragm spring to open the damper. On Ford and American Motors mechanically-operated units, an auxiliary air inlet valve and the vacuum motor that operates it, are mounted on the rear of the air cleaner housing. The sudden drop in manifold vacuum that accompanies quick throttle opening, permits the valve to open and provide the additional air required for acceptable acceleration. Earlier Ford models employed a vacuum override motor (mounted beneath the air cleaner snorkel) that momentarily opened the valve plate by the action of a piston rod.

Air entering the carburetor, either from the heat stove or from the underhood area, passes through the air cleaner element where the dust and dirt particles contained in the air are removed.

On dual-snorkel heated air cleaners, one snorkel is connected to the temperature sensor and draws air through the heat stove exactly as does the single snorkel air cleaner previously described. The other snorkel functions strictly by manifold vacuum. The two vacuum motors are connected by a "Tee" in the vacuum hose connected to the carburetor. During periods of hard acceleration, manifold vacuum drop permits both vacuum motors to open the snorkels to underhood air permitting sufficient air intake for maximum performance. During periods of high-speed operation, the increased air velocity causes a pressure drop inside the air cleaner. With a higher pressure on the outside of the door, the pressure differential tends to open both snorkel doors to underhood air regardless of manifold vacuum.

On many late-model cars the application of heated air to the carburetor on cold engine starts has also lead to the elimination fo the exhaust manifold heat control valve and its sticking problems.

IDLE STOP SOLENOIDS

IDLE STOP SOLENOIDS

GENERAL MOTORS AND CHRYSLER TYPE

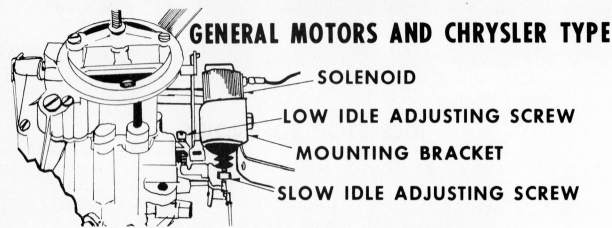

- SOLENOID
- LOW IDLE ADJUSTING SCREW
- MOUNTING BRACKET
- SLOW IDLE ADJUSTING SCREW

PHASES OF SOLENOID OPERATION

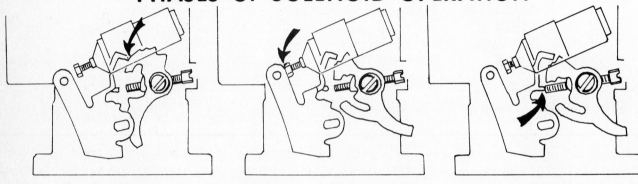

FAST IDLE   **SLOW (CURB) IDLE**   **LOW (SHUTDOWN) IDLE**

FORD SOLENOID THROTTLE MODULATOR

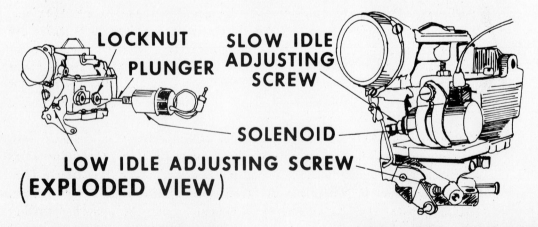

LOCKNUT
PLUNGER

SLOW IDLE ADJUSTING SCREW

SOLENOID

LOW IDLE ADJUSTING SCREW

(EXPLODED VIEW)

USED ON AUTOLITE CARBURETOR   **USED ON CARTER CARBURETOR**

IDLE STOP SOLENOIDS

The idle stop (or anti-Dieseling) solenoid is used on the engines of many late model vehicles.

Most exhaust emission controlled engines idle at higher speeds, use leaner fuel mixtures, with retarded ignition timing, and have higher than average cooling system temperatures than engines that are not emission controlled. Further, the idle speed of the engine is set with the transmission selector in DRIVE position. When the driver places the selector in NEUTRAL, before shutting the engine OFF, the idle rpm increases when the transmission load is removed. All these factors increase the tendency of the engine to "run on" or "Diesel" after the ignition switch has been turned OFF. Because the engine keeps running on only a few of its cylinders, run-on is usually very rough and is frequently associated with violent detonation. The function of the idle stop solenoid is to prevent this tendency by permitting the idle speed to drop low enough to virtually shut off the engine's air supply thus causing the engine to stop running immediately when the ignition is turned OFF.

GENERAL MOTORS AND CHRYSLER SOLENOID

The idle stop solenoid is bracket-mounted on the carburetor and works in the following manner. When the ignition switch is turned ON, the solenoid plunger is electrically activated and moves to an extended position. This action places the adjusting screw the plunger carries, in contact with the throttle linkage where it serves as the idle speed control. When the ignition switch is turned OFF, the plunger is deactivated and retracts into its housing. Since the plunger adjusting screw now ceases to serve as the throttle stop, the throttle linkage moves to a lower idle setting established by the carburetor idle speed adjusting screw.

On the Chrysler solenoid the adjusting screw is mounted on the linkage contacting the solenoid plunger rather than on the plunger itself. The idle adjusting screw and solenoid further serve as a switch to activate the distributor solenoid, on engines so equipped. The function of the distributor solenoid is to provide spark retard during idle and closed throttle deceleration periods. This subject is covered in the Distributor Section of this manual.

The chart illustrates the three phases of solenoid operation. When a cold engine is started it runs at a fast idle on partial choke. The solenoid is activated and the plunger is extended but it cannot reach the throttle linkage because the fast idle cam is keeping the engine running at a speed higher than the one for which the solenoid is set. At normal operating temperature (choke open) the solenoid plunger governs the engine idle speed by acting as the throttle stop. The plunger remains extended during all periods of engine operation. With the ignition switch turned OFF and the engine shutting down, the solenoid is deactivated and the plunger draws back into the solenoid housing. The engine rpm is now controlled by the carburetor slow idle adjusting screw. This speed is low enough to prevent "run on" but just high enough to prevent the throttle plate(s) from closing completely and scuffing the throttle bore(s).

On warm engine start-up there will be, of course, no fast idle cam action. The solenoid plunger will immediately serve as the throttle stop. But here is an important point. The solenoid, though activated when the ignition switch is turned ON, may not be strong enough to open the closed throttle against the weight of the linkage and the action of the return spring. After turning on the switch it is necessary to depress the accelerator approximately one-third of its travel to permit the plunger to move to its extended position where it will then be able to hold the throttle open. Failure of the engine to continue running after start-up may be experienced if this procedure is not observed.

When starting a cold engine, the accelerator must be fully depressed to set the choke and fast idle cam as well as the solenoid.

FORD MOTORS THROTTLE MODULATOR

When used on Ford-built 6-cylinder engines, the idle stop device is called a Solenoid Throttle Modulator. It serves the same function and operates in the same manner as previously described.

The throttle modulator used on the Carter YF carburetor is adjusted for setting the slow (curb) idle by turning the adjusting screw as mentioned above. The solenoid used on the Autolite carburetor is adjusted by loosening the locknut and turning the solenoid in or out to obtain the specified slow idle and then tightening the locknut. The solenoid on some late-model engines is mounted on an adjustable bracket slide providing a means of setting the solenoid adjustment with the air cleaner installed.

When setting the low (shutdown) idle on Ford solenoid throttle modulators, the solenoid is electrically disconnected at the bullet connector near the loom, not at the solenoid, as is the General Motors unit.

THREE IDLE SPEEDS

With the introduction of the idle stop solenoid (or throttle modulator) there also came the third idle speed setting. Now there is a fast idle; a slow (curb) idle; and a low (shut-down) idle.

The cold fast idle is set by a throttle linkage fast idle adjusting screw (or by bending a fast idle cam follower) to a specified speed which will be high enough to keep a cold, partially-choked engine running without stalling or loading.

The slow (curb) idle is set by adjusting the solenoid plunger adjusting screw (GM & Ford); by setting the adjusting screw contacting the solenoid plunger (Chrysler); by turning the solenoid or adjusting the solenoid bracket (Ford) to the recommended idle speed setting.

The low (shut-down) idle is set by adjusting the carburetor low idle adjusting screw after electrically disconnecting the solenoid lead at the solenoid

(GM) or at the bullet connector at the loom (Ford). The Chrysler low idle is set with the solenoid energized and the engine idling by adjusting the carburetor idle speed screw inward until the tip of the screw just touches the stop on the carburetor. Then back the screw out one full turn to obtain a slow idle of approximately the desired setting.

The three idle speeds may be for example: a fast idle of 2000 rpm; a slow idle of 750 rpm; and a low idle of 500 rpm. Setting all three idle speeds to specifications is an important part of the tune-up procedure.

DECELERATION VALVE

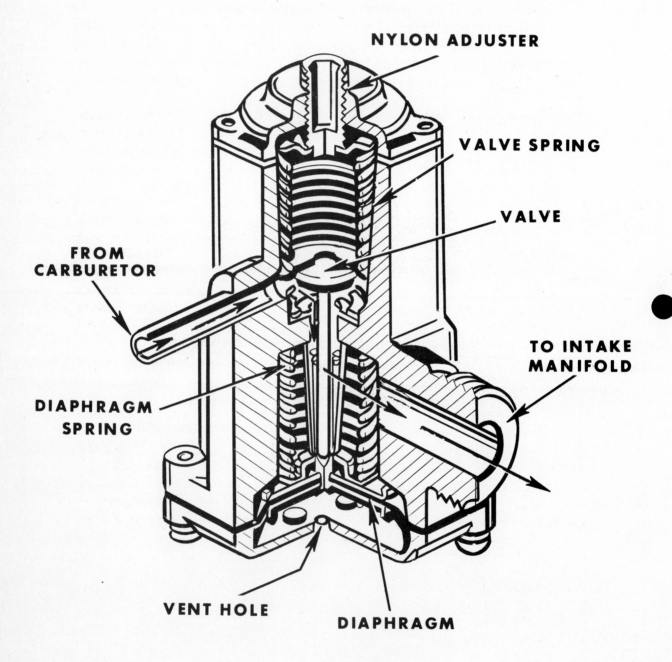

NYLON ADJUSTER

VALVE SPRING

VALVE

FROM CARBURETOR

TO INTAKE MANIFOLD

DIAPHRAGM SPRING

VENT HOLE

DIAPHRAGM

DECELERATION (DECEL) VALVE

Ford's 1600cc and 2000cc engines used to power the Pinto and Capri cars are fitted with an assist emission control device called a Deceleration Valve. The valve was introduced on some 1971 models.

The generation of objectionable hydrocarbon and carbon monoxide exhaust emissions are greatest during periods of deceleration. The function of the deceleration valve is to assist in controlling these emissions by metering additional amounts of fuel and air into the engine during these periods.

The vacuum-operated deceleration valve is mounted on the intake manifold adjacent to the carburetor. The valve housing contains a spring-loaded valve in its upper chamber and a spring-loaded diaphragm in its lower chamber. The bottom cover contains a bleed hole to permit the constant entry of atmospheric pressure to the lower side of the diaphragm.

A nylon or plastic adjusting screw positioned in the top cover provides a means of setting spring tension that governs both the valve opening time and its duration of opening.

One hose connects the deceleration valve to the "deceleration section" of the carburetor. This section is cast in the carburetor's upper housing and consists of a fuel pick-up tube and an air bleed tube. Both tubes have precision drilled restrictors and a common outlet tube to which the deceleration valve hose is connected. Another hose connects the valve to the intake manifold.

Only during periods of engine deceleration is the intake manifold vacuum strong enough to lift the deceleration valve diaphragm against spring tension thus lifting the valve from its seat. With the valve open, intake manifold vacuum draws a metered amount of fuel and air from the carburetor deceleration section. The air/fuel mixture passes through the valve and into the intake manifold. This additional fuel and air provides a greater volume of conbustible mixture in the engine cylinders, producing a more rapid and complete combustion process. Reduced levels of exhaust emissions are thereby effected.

The period of additional fuel feed time is calibrated to last for about 3 to 5 seconds after which time the combined action of the drop in manifold vacuum and the tension of the diaphragm spring and the valve spring, closes the valve.

EXHAUST GAS RECIRCULATION (EGR) SYSTEMS

CONTROLLED EGR SYSTEM

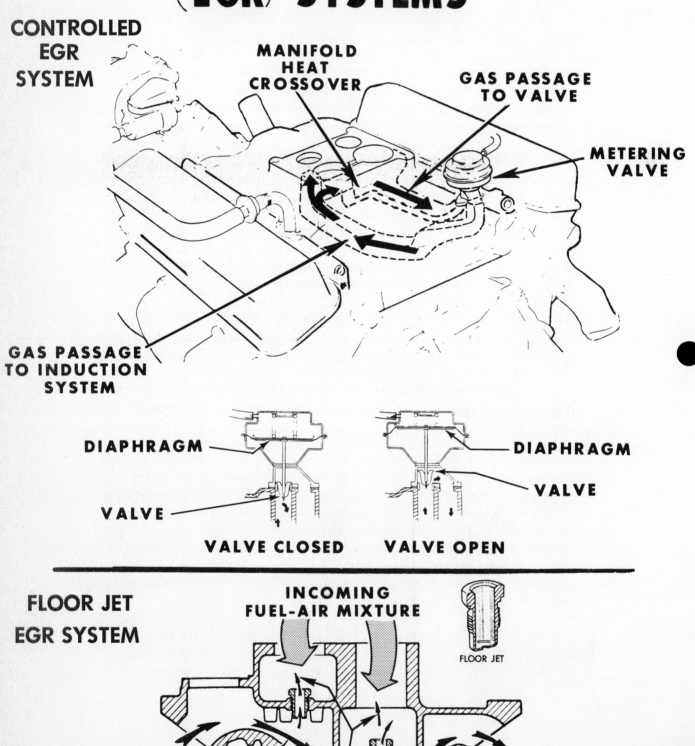

MANIFOLD HEAT CROSSOVER

GAS PASSAGE TO VALVE

METERING VALVE

GAS PASSAGE TO INDUCTION SYSTEM

DIAPHRAGM

DIAPHRAGM

VALVE

VALVE

VALVE CLOSED

VALVE OPEN

FLOOR JET EGR SYSTEM

INCOMING FUEL-AIR MIXTURE

FLOOR JET

EXHAUST GAS CROSS-OVER

RECIRCULATING GASES

EXHAUST GAS RECIRCULATION (EGR) SYSTEMS

The Exhaust Gas Recirculation Systems introduced on some 1972 car models are designed to control the formation of the various oxides of nitrogen (NOX) in exhaust gases.

The high temperatures and lean air/fuel ratios that are instrumental in the control of hydrocarbon (HC) and carbon monoxide (CO) emissions, unfortunately are conducive to increasing the formation of nitrogen oxide emissions. It therefore became necessary to design a system that would effectively limit the formation of oxides of nitrogen without unfavorably influencing the control of HC and CO emissions. The Exhaust Gas Recirculation System serves this important function.

The EGR System is based on the principle that the formation of oxide of nitrogen emissions can be limited by reducing the peak temperatures of the burning fuel charge in the engine's combustion chambers. This can be accomplished by the introduction of an inert material into the fresh fuel charge and since exhaust gas is an inert material, and is in plentiful supply, it is used in the EGR System for induction into the intake manifold.

Diluting the fresh fuel charge with a calibrated amount of exhaust gas was formerly accomplished by the engine valve timing pattern that increased the degree of valve overlap. By having both the intake valve and exhaust valve open in the same cylinder at the same time, for a limited number of degrees, the intake charge was somewhat diluted by the last of the exhausting gases leaving the cylinder. When the need for more fuel charge dilution than the valve overlap period could provide became a requirement, the Exhaust Gas Recirculation System was devised.

CONTROLLED EXHAUST GAS RECIRCULATION SYSTEM

This Exhaust Gas Recirculation System is composed of a valve and two passages cast in the intake manifold. One passage leads from the manifold heat crossover passage to a metering valve. The other passage leads from the valve to centrally-positioned holes in the intake manifold floor below the carburetor.

The metering valve is normally held closed by a coil spring positioned above the valve diaphragm. Diaphragm movement is under the influence of a vacuum signal initiated at a point above the carburetor closed throttle plates. At idle speed, the very weak carburetor vacuum cannot overcome the diaphragm spring tension and the valve remains closed. Dilution of the fuel charge is prevented and a smoother idle is maintained. Since the formation of nitrogen oxides is lowest at idle speed, dilution of the fuel charge is not essential at idle.

As the throttle is opened, the formation of nitrogen oxides increases rapidly. To control this condition, the opening throttle plates expose the metering valve diaphragm to increased intake manifold normal vacuum, the valve is lifted off its seat, and a metered amount of exhaust gas is permitted to enter the induction system to limit the formation of NOX emissions, precisely when needed.

The controlled EGR valve can be checked to determine if the valve opens and closes properly. Procedures vary, but the general method is to operate the engine at a fast idle and then remove and reconnect the valve's vacuum line while noting the change in speed. When the vacuum line is off, the speed should increase, typically, about 50–100 rpm. The speed should return to its original speed when the vacuum line is reconnected. This indicates that the valve closes (speed increases) when the vacuum is interrupted, and opens (speed decreases) when normal carburetor vacuum is applied to the valve. If the test speed is too low, normal EGR or carburetor vacuum will not be present and the test will be inconclusive. At idle speeds, no vacuum is applied to the EGR valve thereby keeping it closed.

This EGR System is classified as "Variable" since the control valve action determines when and to what extent fuel charge dilution occurs.

FLOOR JET EXHAUST GAS RECIRCULATION SYSTEM

This type of EGR System provides for the entrance of exhaust gas into the intake manifold of the V-8 engines through metered orifice jets positioned in the floor of the manifold heat crossover passage below the carburetor. The orifice jet in the 6-cylinder engines is located in the manifold "hotspot" below the carburetor heat riser.

The calibrated size of the orifice in each jet determines the amount of exhaust gas that can be inducted into the intake manifold by engine vacuum.

The jets are made of nonmagnetic steel and are threaded into the floor of the intake manifold so that they can be removed for cleaning of the orifice, should the need occur.

CHECKING EGR SYSTEMS

This EGR System is classified as "Fixed" since the amount of fuel dilution is controlled by the fixed size of the orifice in the jet.

The floor jet EGR system can be checked only by visual examination after removing the carburetor. If the floor jets, located below the carburetor, appear clogged or otherwise damaged, they must be replaced. They must be removed carefully since, being made of stainless steel, they are non-magnetic and difficult to retrieve if dropped.

120

EXHAUST GAS RECIRCULATION
(EGR) SYSTEMS

DUAL DIAPHRAGM EGR VALVE

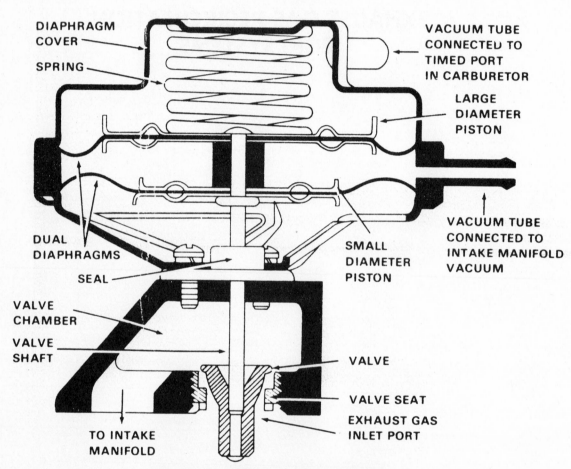

**DUAL DIAPHRAGM
EXHAUST GAS RECIRCULATION VALVE**
(PARTIALLY OPEN-CRUISE)

DUAL DIAPHRAGM EGR VALVE

A modified version of the EGR valve shown previously will also be found. This is the dual diaphragm or proportional type valve. Its purpose is to provide more precise control over the amount of exhaust gas admitted to the incoming mixture so as to provide improved engine operation and driveability.

In principle, it is similar to the single diaphragm valve. However, the additional diaphragm makes the valve more responsive to changes in intake manifold vacuum. These valves have two vacuum lines connecting to it. One goes to the vacuum port on the carburetor, the same as on the single-diaphragm valve. A second vacuum line connects to the intake manifold.

FUEL EVAPORATION EMISSION CONTROL SYSTEMS

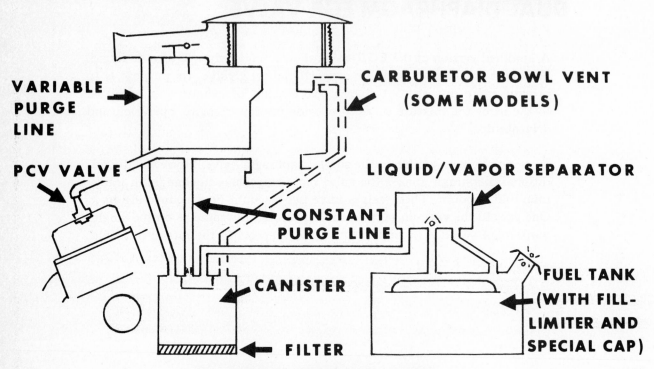

VARIABLE PURGE LINE

CARBURETOR BOWL VENT (SOME MODELS)

PCV VALVE

LIQUID/VAPOR SEPARATOR

CONSTANT PURGE LINE

CANISTER

FUEL TANK (WITH FILL-LIMITER AND SPECIAL CAP)

FILTER

CANISTER STORAGE TYPE

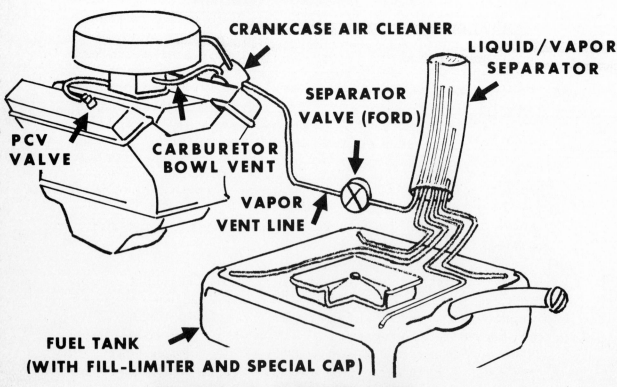

CRANKCASE AIR CLEANER

LIQUID/VAPOR SEPARATOR

SEPARATOR VALVE (FORD)

PCV VALVE

CARBURETOR BOWL VENT

VAPOR VENT LINE

FUEL TANK (WITH FILL-LIMITER AND SPECIAL CAP)

CRANKCASE STORAGE TYPE

FUEL EVAPORATION EMISSION CONTROL SYSTEMS

The Fuel Evaporation Emission Control System was developed and designed to prevent the escape of gasoline vapors from the fuel tank and carburetor into the atmosphere. This system is in addition to the Positive Crankcase Ventilation System and the various Exhaust Emission Control Systems.

Starting with the 1972 car models, all vehicles will employ the popular system of storing the fuel vapors in a charcoal granule-filled canister. Some earlier models also used this system while others employed the engine crankcase as the storage area for the fuel vapors.

Essentially the Evaporation Emission Control Systems function in the following manner. When the engine is running, the fuel vapors are conducted through tubing from the fuel tank to the carburetor or carburetor air cleaner to be immediately consumed in the engine. When the engine is not running, the tank and carburetor fuel vapors are either piped into a charcoal granule-filled canister or they are stored in the engine crankcase, depending on system design. When the engine is started air passing through the canister, or through the crankcase ventilation system, picks up the fuel vapors and carries them into the engine where they are consumed.

By this method of control, gasoline vapors can escape only through the Evaporation Control System thereby eliminating fuel vapors as a source of air pollution.

VAPOR TRANSFER SYSTEM

The fuel tank cap is either nonvented or is a pressure/vacuum sensitive type thereby preventing the escape of either liquid gasoline or gasoline vapors through the tank filler. To assure that the tank will be constantly vented regardless of the vehicle's attitude, two, three or four points are used as tank vents. Lines from these tank vent points meet at a unit called a liquid/vapor separator which is positioned just forward of, and slightly above, the fuel tank. Only fuel vapors can pass the separator and enter the single vapor transfer line leading forward to the canister or crankcase. Liquid fuel is returned to the tank through a return vent line.

The separator front vent section is equipped with a needle-and-seat (or ball-and-seat) positive shut-off valve. In the event of vehicle upset, liquid gasoline cannot flow uncontrolled into the canister or crankcase. Because of fuel tank position on some station wagons, a separator is not used on these vehicles.

On some vapor control systems a two-way pressure-vacuum relief valve is positioned just forward of the separator. The valve's function is to prevent fuel starvation or mechanical damage to the fuel tank. If fuel tank pressure falls too low, or rises too high, the valve will admit air or expel vapor through the transfer line thereby restoring normal pressure values. On some systems the valve maintains a positive pressure of 1 psi in the fuel tank and vent system to aid in retarding further fuel evaporation.

CHARCOAL CANISTER STORAGE

From the relief valve (or separator) the vapor transfer line runs to an activated charcoal-containing canister which is located in the engine compartment. Another line leads from the canister to the carburetor PCV hose. As previously stated, when the engine is running the tank fuel vapors are constantly being drawn into the engine. When the engine is shut down, the fuel vapors generated by "heat soak" are absorbed into the activated charcoal. When the engine is restarted, engine vacuum draws air through the canister filter, over the charcoal granules removing the absorbed fuel vapors and drawing them into the engine.

Some vapor control systems employ two vent lines from the canister. One line runs from the canister to the carburetor PCV hose, previously mentioned. This line is called the "Constant Flow Purge Line". Its function is to permit the engine manifold vacuum to draw a constant uniform amount of fuel vapor from the canister at all engine speeds and operating conditions including engine idle periods. Vapor flow is controlled by a fixed orifice positioned in the top of the canister. The second line runs from the canister to the carburetor air cleaner snorkel. This line is usually called the "Variable Flow Purge Line". The vapors that pass into the engine through this line are in variable amounts depending on engine speed, particularly at speeds above idle. This arrangement permits complete canister purge at cruising speeds.

The activated charcoal granules contained in the canister have a capacity of approximately 50 grams of vapor which is the equivalent of between 2 and 3 ounces of liquid gasoline. This capacity is adequate to effectively contain fuel vapors even when the vehicle is parked for an extended period. The amount of fuel that passes through the vapor control system at any time is too small to have any measurable effect on either engine operation or fuel economy.

CRANKCASE STORAGE

All 1970-71 engines built by Chrysler Corporation and some Ford-built engines of the same years used the engine crankcase as the fuel vapor storage area. Since fuel vapors are from 2 to 4 times heavier than air, they settle in the crankcase on top of the oil. When the engine is running, the fuel vapors are purged from the crankcase and drawn through the Crankcase Ventilation System into the engine. On the Ford engine a combination valve placed in the vapor line serves to isolate the fuel tank from engine induced pressures.

CARBURETOR VAPOR CONTROL

The control of carburetor fuel vapor is particularly effective during periods of "heat soak" which occur immediately after engine shut-down. These are the periods of maximum carburetor fuel vapor emission.

The method of carburetor vapor control consists of a line from the carburetor bowl to the crankcase on some installations or on some 4-barrel carburetors, a line connecting the carburetor primary fuel bowl to the vapor canister.

In other installations, the generation of fuel vapors are somewhat minimized by the use of an aluminum heat-dissipating plate which contains the carburetor-to-intake manifold gasket. The plate serves as a heat shield to deflect and dissipate the engine heat to the surrounding air thereby maintaining a relatively low carburetor fuel temperature. The possibility of fuel boil-away is consequently considerably relieved.

FUEL TANK

A slight redesigning of the upper portion of the fuel tank is also a part of the Vapor Emission Control System. An over-fill protector or fill-limiter (an inverted dish-like member) is mounted in the top of the fuel tank. In some installations this member is actually a small separate tank with a capacity of slightly more than one gallon and is connected to the main tank by small metered holes. Regardless of its size or shape, the limiter is designed to remain essentially empty after the fuel tank is filled. A fuel tank top area collection space is thereby maintained for the fuel vapors and an expansion space for increased fuel volume. These conditions occur when a filled fuel tank is subjected to high temperature as when the vehicle is parked in the hot sun.

SERVICE

Service of the Fuel Evaporation Emission Control System is generally limited to cleaning or replacing the canister filter at 12,000 mile intervals. The separator is a sealed unit and requires no service or maintenance.

When an engine equipped with a Vapor Control System is being tuned, the fuel tank vent line should be disconnected at the canister. The line should not be plugged. Disconnecting the line prevents fuel vapors from being drawn into the engine and upsetting final adjustments.

THE CHARGING CIRCUIT

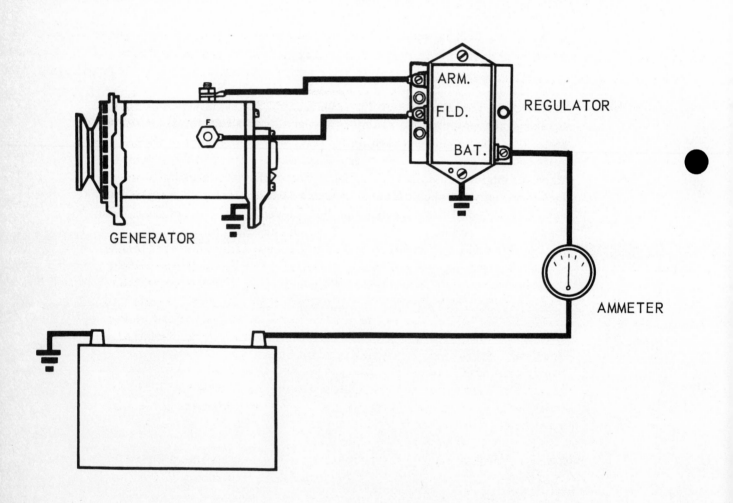

GENERATOR

ARM.

FLD.

BAT.

REGULATOR

AMMETER

THE CHARGING CIRCUIT

The purpose of the charging system is to supply current for the lights, ignition, radio, heater and other electrical accessories and to maintain the storage battery in a charged condition or to recharge it when it has become discharged. The charging system consists of four important components: battery, generator or alternator, regulator and leads.

The battery performs two important functions. First, it supplies electrical energy for the starting motor and the ignition system until the engine starts. Second, it intermittently supplies current for the lights, radio, heater and other accessories when the electrical demands exceed the output of the generator or alternator.

The generator or alternator converts mechanical energy supplied by the engine into electrical energy. This energy is used to charge the battery and to supply power to the electrical system of the automobile.

The regulator is the control unit for the generator or alternator. Its function is to automatically control generator output to safely meet all conditions of speed and electrical load. The generator regulator assembly usually consists of three units, a cutout or reverse current relay, a voltage limiter and a current regulator. The alternator regulator contains a voltage control unit and may also contain a field relay and/or an indicator light relay.

The cutout relay or circuit breaker is used to open and close the circuit between the generator and battery. When the generator voltage exceeds the battery terminal voltage, the contact points close. This completes the circuit between the generator and battery permitting the generator to charge the battery. If the engine is stopped, is idled or operated at low speed, the contact points open and prevent the battery from discharging through the generator.

The voltage limiter is used to regulate the generator voltage to a predetermined value, thus protecting the battery and all vehicle electrical accessories from the damage of high voltage.

When the generator voltage rises to a value at which the voltage limiter is set, the regulator contacts vibrate thus inserting a resistance in the generator field circuit thereby limiting the generator voltage.

The current regulator is used to limit the generator output to a predetermined value, thus protecting the generator against overload. When the generator output reaches its predetermined safe maximum output, the regulator contacts vibrate inserting a resistance in the field circuit which reduces the voltage and consequently the amperage output of the generator.

The field relay is used to energize the alternator field coil by connecting the coil to the battery when the ignition switch is turned on. When the ignition switch is turned off, the field relay opens the circuit between the field coil and the battery to prevent battery discharge.

The indicator light relay is used in the alternator-equipped system where a light is mounted on the instrument panel to replace the ammeter. The relay permits the light to go on when the alternator is not producing voltage and causes the light to go out when voltage is produced. Should the light be lit when the engine is running, trouble is indicated.

The leads are necessary to complete the circuitry between the charging system components and the vehicle. The wiring insulation must be in good condition. The connections must be clean and tight.

All components in a charging system are equally important. In order for the charging system to function properly, the generator must be capable of producing an output at least equal to its electrical rating. The regulator in turn must be able to limit the generator output at a predetermined setting. The leads should be in good condition and the connections clean and tight. With high resistance in the leads and connections the generator and regulator cannot perform properly.

GENERATOR OPERATING PRINCIPLES

GENERATOR OPERATING PRINCIPLES

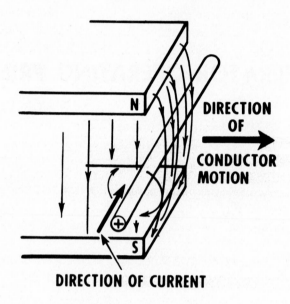

DIRECTION
OF
CONDUCTOR
MOTION

DIRECTION OF CURRENT

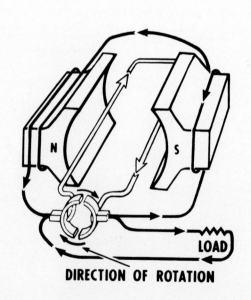

LOAD

DIRECTION OF ROTATION

GENERATOR OPERATING PRINCIPLES

Current is developed by the automobile generator by moving wire conductors, in the form of an armature, through a magnetic field generated by current flow through field coils which are wrapped around pole shoes.

When a conductor is moved through a magnetic field, it cuts the magnetic lines of force of the field and induces a voltage in the conductor causing current to flow in the conductor. The direction in which the current will flow in the conductor is governed by the direction of the movement of the conductor in the magnetic field. As the conductor moves through the field it causes a distortion of the lines of magnetic force which tends to cause the lines of force to wrap themselves around the conductor, so to speak. Applying the right hand rule to this distorted field movement will verify the direction of current flow in the conductor.

Illustrated are the basic elements of a generator circuit. The armature wire conductor is illustrated as a single loop of wire. Each end of this loop is connected to commutator segments. The function of the commutator is to rectify the alternating current generated by the revolving armature loop into direct current.

As the wire loop is rotated in the magnetic field, voltage is induced in the loop causing current to flow as previously explained. The current generated flows from the insulated brush through the connected electrical load circuit and back to the generator ground brush, completing the circuit.

A portion of the generator output is shunted through the field coils to develop and sustain the magnetic field in which the armature turns. Generators are identified as "shunt wound" because the field is shunted across the armature.

As the armature is turned faster, more lines of force are cut resulting in increased generator output. This greater output also results in increased current flow through the field coils. This increases the strength of the magnetic field and consequently increases, still farther, generator output. Unless this action is controlled, generator output will continue to increase to the point where so much current and heat will be generated that the soldered connections of the armature windings will melt and the generator will be destroyed. To control this condition, a regulator is used which governs generator output by controlling the strength of the magnetic field.

GENERATOR FUNCTION

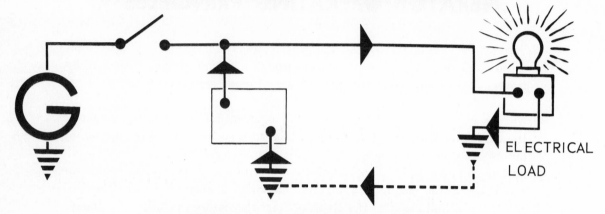

BATTERY POWER ONLY

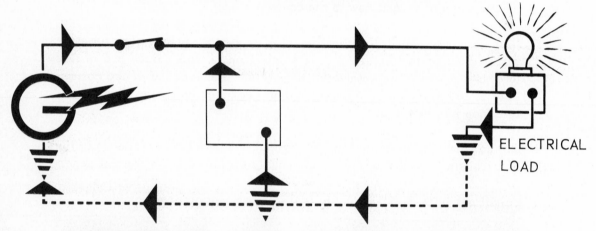

GENERATOR AND BATTERY POWER

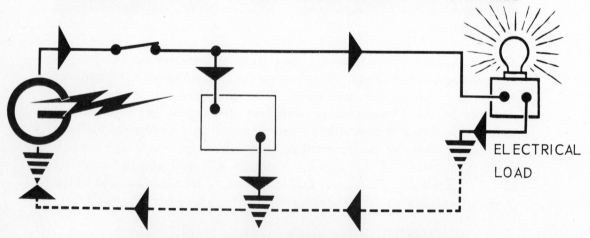

GENERATOR POWER ONLY

GENERATOR FUNCTION

With the generator at rest, or operating at low speed, all of the electrical energy is furnished by the battery.

Whenever the electrical load exceeds the output of the generator, then electrical energy is also furnished by the battery to supplement the generator output.

When an electrical load is less than the generator output, the generator supplies all the power for the load and also recharges the battery.

The generator on a vehicle has been carefully selected so that it will meet all the electrical requirements of that vehicle. Its rated output will be high enough to efficiently supply the electrical load of the vehicle and its accessories at varying speeds.

The principle advantage of the alternator over the generator is that the alternator has an output at engine idle which the generator does not have and the alternator also has a higher low-speed output, if it is belted with correct pulley ratios.

GENERATOR TESTING

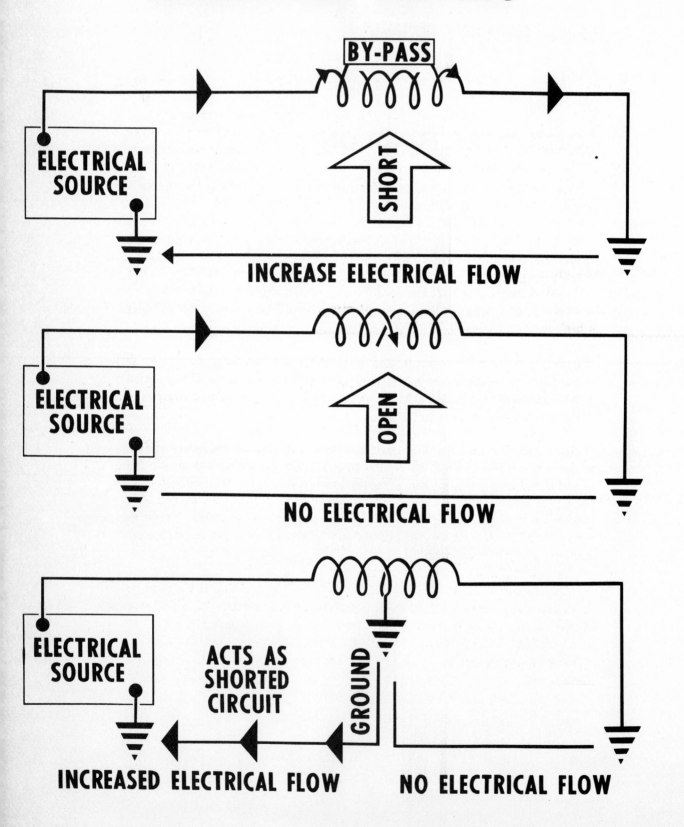

BY-PASS

ELECTRICAL SOURCE

SHORT

INCREASE ELECTRICAL FLOW

ELECTRICAL SOURCE

OPEN

NO ELECTRICAL FLOW

ELECTRICAL SOURCE

ACTS AS SHORTED CIRCUIT

GROUND

INCREASED ELECTRICAL FLOW

NO ELECTRICAL FLOW

GENERATOR TESTING

The charging system like any other system or component on the vehicle, requires periodic service or maintenance to assure top operating efficiency. Should trouble occur in this system, the cause and location of this trouble can readily be determined through a systematic test procedure. To efficiently test charging system components so that proper service operations be performed, it is necessary to have a clear understanding of electrical troubles. Basic malfunctions can be classified into four groups. SHORT circuits, OPEN circuits, GROUNDED circuits and circuits with abnormally HIGH or LOW RESISTANCE.

A short circuit is any accidental contact that permits the current to bypass a portion of the electrical circuit. This condition is present when one or more windings on a coil are bypassed due to insulation failure. A short circuit results in a lower than normal circuit resistance thereby permitting a higher than normal current flow.

An open circuit is an undesired break in the circuit. A break can occur in any one of a number of locations in the circuit such as, coil windings, wires or connections, resulting in an inoperative circuit. No current will flow through an open circuit.

A grounded circuit is an undesired connection that bypasses part or all of the electrical units, from the insulated side to the ground side of the circuit. In a lighting circuit, for example, should a ground occur between the battery and the lamps, the load on the battery would become unreasonably high while the lamps would fail to light. In general, a grounded circuit results in a higher than normal current flow produced by a proportionate reduction in circuit resistance.

A circuit with abnormally high resistance is one containing resistance of a nature that increases the total resistance of the circuit. Poor or loose connections, corroded connections, and frayed or damaged wires are examples of conditions causing high resistance. Should this condition exist, current flow will be less than normal because of the increase in circuit resistance.

CUTOUT RELAY

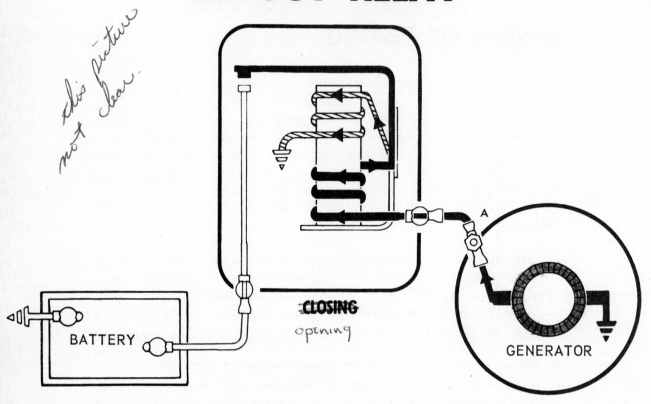

CLOSING

opening

BATTERY

GENERATOR

A

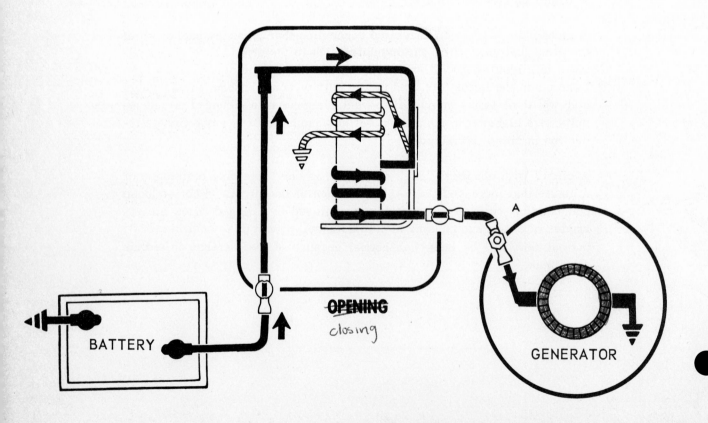

OPENING

closing

BATTERY

GENERATOR

A

CUTOUT RELAY

The cutout relay is an electromagnetic switch used to close the charging circuit between the generator and the battery when the generator voltage is higher than battery voltage and to open the circuit when battery voltage is higher than generator voltage. Without the cutout relay the battery would discharge through the generator to ground when the generator was slowed down or stopped. Such a discharge would damage the generator and permit the battery to become discharged.

The cutout relay has a coil of fine wire connected in shunt from the generator output lead to ground. This coil is a voltage sensitive winding. The generator voltage creates a magnetic field in this coil of the relay. When sufficient generator voltage is available, it overcomes the spring tension and closes the contact points. Generator current can now flow to the battery. After the contacts have closed, regardless of whether the current flows from the battery or the generator, the coil would maintain the same magnetic polarity. Under these conditions the cutout would not be able to open because the shunt winding would keep the relay energized at all times. To correct this condition, another winding, a current sensitive winding, is added to the relay in series with the generator and battery.

With the generator charging, current is flowing in the same direction in both the series and shunt windings. The magnetic field of both windings combines to hold the cutout points closed. When the generator is not charging, the current reverses in the series winding because the battery voltage is higher. Since the current flow in the windings is now in opposite directions, the magnetic fields cancel each other out, the relay becomes demagnetized, and the spring opens the contact points.

A cutout relay is tested to determine whether it closes at the proper generator voltage, and opens at the proper reverse current value. Unless the cutout relay is operating correctly, a discharged battery and/or damage to the charging system can result.

VOLTAGE REGULATOR

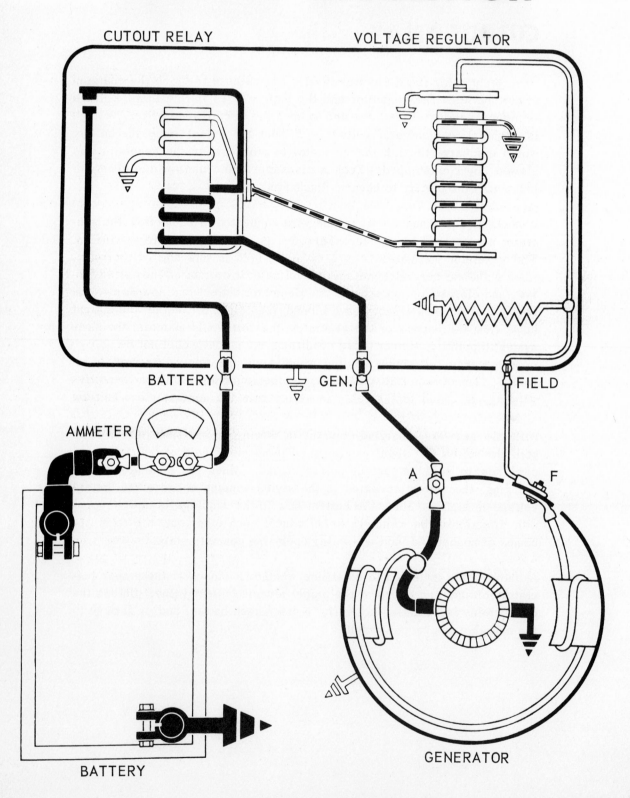

CUTOUT RELAY

VOLTAGE REGULATOR

BATTERY

GEN.

FIELD

AMMETER

A

F

BATTERY

GENERATOR

VOLTAGE REGULATOR

The voltage regulator unit is used to prevent the generator from developing an excessively high voltage. The voltage regulator limits the charging voltage to a value which is safe for the electrical accessories and is able to maintain the battery in a state of full charge.

The voltage regulator contacts are installed in series with the field winding. All of the field coil current must flow through the points of the voltage regulator to ground. As long as the voltage regulator points are closed, the field current and the generator output will attain a maximum value for any specific generator speed. The fine wire coil of the voltage regulator relay is connected across the generator output circuit enabling it to sense the output voltage. As the voltage output of the generator reaches the safe maximum limit, the magnetic field of the voltage sensitive regulator coil becomes stronger and will overcome the spring tension and open the points. With the points open, the field coil circuit is then completed through a resistor which lowers the field current flow thereby decreasing generator output. The decreased voltage output of the generator reduces magnetic strength in the voltage regulator coil and the spring closes the contacts completing the field circuit to ground which allows the generator output to rise. The points vibrate at a high frequency of from 50 to 250 cycles per second thereby regulating the generator output voltage to a specified setting.

The voltage regulator setting is controlled by adjusting the spring tension on the regulator armature. This tension must be carefully adjusted to effect unit operation within specified limits. A voltage regulator that permits a high voltage output will cause damage to the electrical units in the vehicle. Also the battery would be overcharged and eventually damaged. A low voltage setting will not permit the generator to bring the battery to its full state of charge also the electrical accessories would not receive the correct voltage for efficient operation.

CURRENT REGULATOR

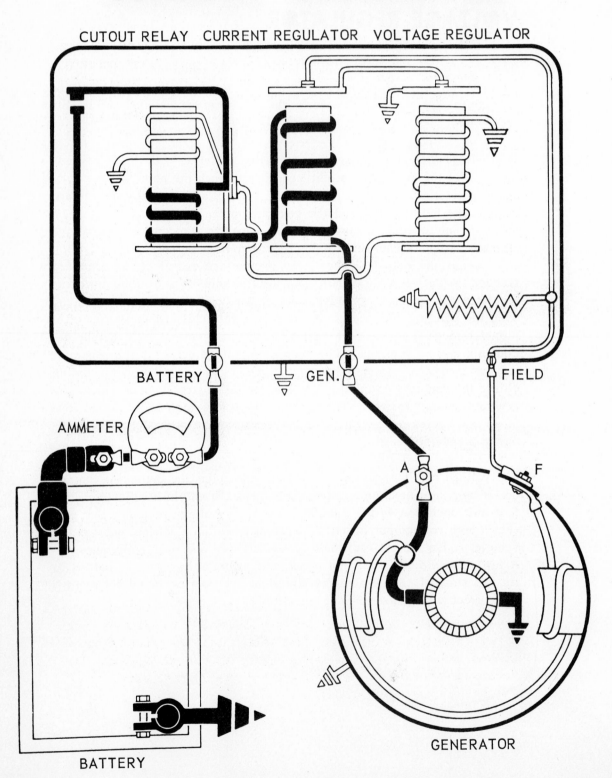

CUTOUT RELAY   CURRENT REGULATOR   VOLTAGE REGULATOR

BATTERY   GEN.   FIELD

AMMETER

A   F

BATTERY

GENERATOR

CURRENT REGULATOR

The current regulator unit protects the generator from excessive current output by limiting the current to a value considered safe for the generator. As the generator output reaches the value for which the current regulator is set, the regulator contact points begin to vibrate. This vibration alternately opens and closes the contact points inserting and removing a resistance in the generator field circuit controlling the voltage and thereby limiting the current output of the generator.

The generator field coils are in series with the current regulator points and the voltage regulator points. The current regulator windings are made of heavy wire and carry the entire output of the generator. When the current output reaches the rated maximum output of the generator, a magnetic field in the heavy windings of the current regulator becomes strong enough to overcome spring tension, opening the contact points which breaks the field circuit. The field current then flows through a resistor to ground reducing the output of the generator. The current flow through the regulator windings also drops, reducing the magnetic field thus allowing the points to close and complete the field circuit to ground. Generator output then rises, the relay magnetic field again opens the points inserting the resistance in the field circuit. This action continues as long as current regulation is required.

The current regulator setting is controlled by adjusting the spring tension on the regulator armature.

Both the voltage regulator and current regulator never operate at the same time. If the electrical load requirements are heavy and the battery is low the system voltage will not be sufficient to cause the voltage regulator to operate. The generator output consequently will increase until it reaches the value for which the current regulator is set at which time the current regulator will operate to protect the generator from overload.

If the electrical load is reduced, or if the battery begins to come up to charge, the system voltage will increase to a value sufficient to cause the voltage regulator to operate. When this happens the generator output is reduced below the value required to operate the current regulator. The current regulator then stops operating and all control is dependent on the operation of the voltage regulator.

DOUBLE CONTACT
VOLTAGE REGULATOR

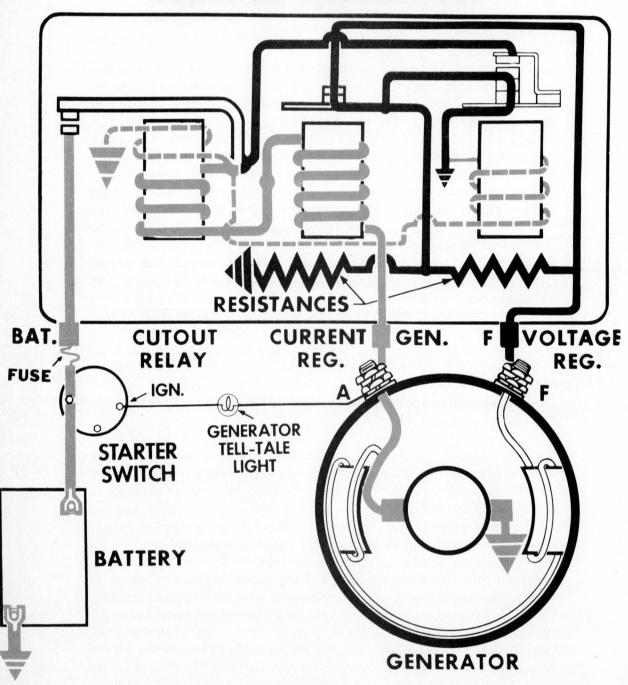

RESISTANCES

BAT.
FUSE

CUTOUT
RELAY

IGN.

STARTER
SWITCH

CURRENT
REG.

A

GENERATOR
TELL-TALE
LIGHT

GEN.

BATTERY

F

VOLTAGE
REG.

F

GENERATOR

DOUBLE CONTACT VOLTAGE REGULATOR

One of the factors that limit generator output is the amount of current that can be carried by vibrating contacts without excessive arcing. The double contact voltage regulator was developed to handle higher generator field current.

When generator speed is low, the first set of contacts on a double contact voltage regulator, control the voltage output in the same manner as in a single contact regulator. When the generator speed increases, these contacts may not effectively control the field current because the generator voltage has become high enough to push sufficient voltage through the resistor to produce a higher than desired output.

When the generator speed and voltage increase to this point, the voltage relay coil will be influenced by an electromagnetic field strong enough to attract the armature to engage the second contact set. These are called the "shorting" contacts. When these points close, the generator field circuit is shorted out. This action practically eliminates all field strength. The generator output drops and the points open. The field current then flows through the resistor. This method of voltage control effectively maintains the generator voltage at a predetermined setting without the tendency of voltage "creep" at high speed.

INDICATOR LAMPS

Many late model cars are equipped with a "tell-tale" indicator lamp in the charging circuit instead of an ammeter.

One side of the indicator lamp is connected to a wire that is "hot" when the ignition switch is turned on and the other side of the lamp is connected to a wire that is "hot" when the generator is charging.

When the ignition switch is turned on, battery current flows through the lamp circuit to ground in the generator causing the lamp to light. When the engine is started generator voltage is applied to this same circuit but to the other side of the lamp. Since approximately equal voltage is applied to both sides of the indicator lamp, no current will flow in the circuit and the lamp goes out.

If the indicator lamp does not light when the ignition switch is turned on, the lamp should be tested and the lamp circuit should be checked for open circuit or loose connections. If the lamp stays on after the engine has started, the generator should be checked for output.

GENERATOR CIRCUITS

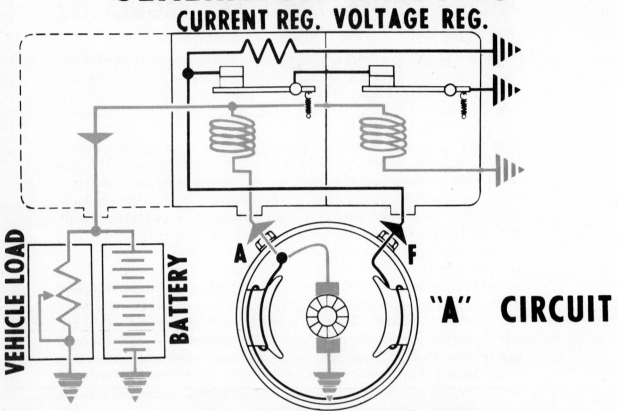

CURRENT REG. VOLTAGE REG.

VEHICLE LOAD

BATTERY

A   F

"A" CIRCUIT

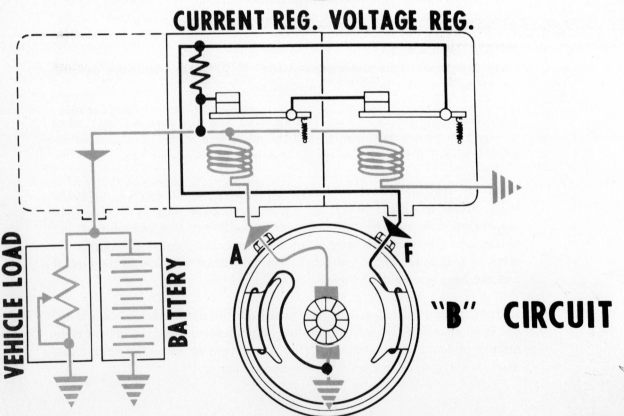

CURRENT REG. VOLTAGE REG.

VEHICLE LOAD

BATTERY

A   F

"B" CIRCUIT

GENERATOR CIRCUITS

There are two types of generator charging circuits; the "A" circuit and the "B" circuit.

In the "A" circuit, current for the generator field circuit starts at the generator insulated brush, flows through the two field coils to the field terminal on the regulator, through the regulator current and voltage regulator points to ground. This field circuit is said to be externally grounded. That is, it is grounded in the regulator. General Motors and Chrysler Corporation frequently use the "A" circuit.

In the "B" circuit, current for the generator field circuit starts in the regulator, flows through the current and voltage regulator points to the field terminal on the regulator, to the generator, through the generator field coils and to ground. This field circuit is said to be internally grounded. That is, it is grounded in the generator. Ford generally uses the "B" circuit.

The construction of the "A" circuit and "B" circuit regulators is similar in many details but they are not interchangeable because of the method of field circuit grounding. When replacing voltage regulators, this factor along with polarity and voltage are essential to correct unit replacement.

To determine if the generator is in an "A" or "B" circuit, perform this simple test:

1. Disconnect the field "F" wire from the **generator**. DO NOT allow this wire to touch ground.

2. Connect a voltmeter from the **generator** field terminal to ground.

3. Operate the engine at a fast idle. If a voltmeter reading is indicated, the generator is in an "A" circuit. If no reading is indicated, the generator is in a "B" circuit.

   Note: In the event the generator is dead, there will be no voltage indicated in the "A" circuit test. An output test will quickly detect a defective generator.

GENERATOR POLARITY

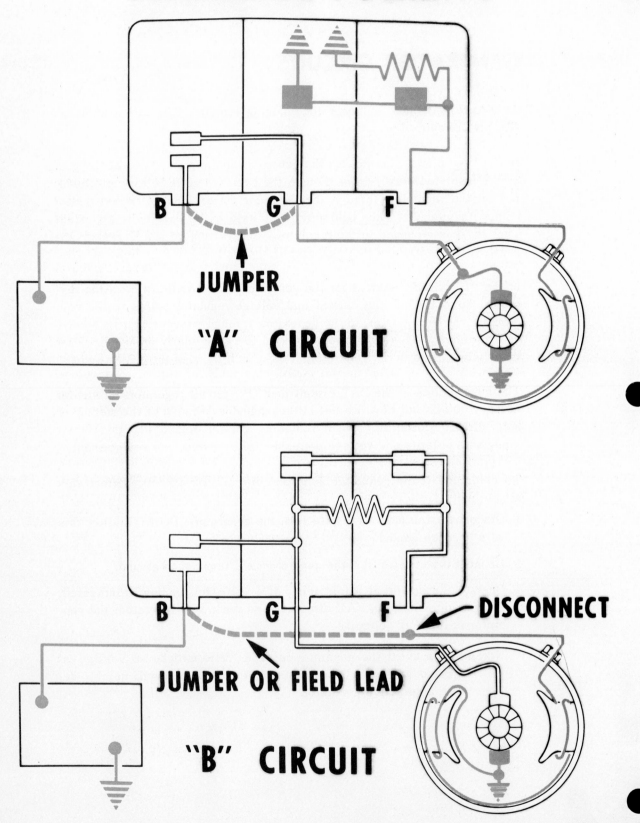

JUMPER

"A" CIRCUIT

DISCONNECT

JUMPER OR FIELD LEAD

"B" CIRCUIT

GENERATOR POLARITY

The generator will build up voltage that will cause current flow in either direction depending upon the polarity of the residual magnetism in the pole shoes which in turn is determined by the direction of current flow in the field coils. The generator polarity must be in agreement with battery polarity in order for current to flow in the proper direction to charge the battery and to prevent damage to the regulator relay points. Reverse polarity causes these points to flutter, arc and burn and can even cause burning of the generator armature and charging system wiring.

Whenever the leads have been disconnected from a generator or after a generator has been repaired, it must be polarized. It is important that the generator be polarized BEFORE starting the engine. This will insure correct polarity and cause current to flow in the proper direction to the battery. An easily accessible place to polarize the generator is at the voltage regulator terminals.

Circuit "A" generators are polarized by momentarily touching a jumper lead from the regulator battery (B) terminal to the regulator armature (ARM) (Gen) terminal, with the engine stopped. A touch of the jumper is all that is required. Battery current will flow through the generator field coils in the right direction to correctly polarize the generator field coil pole shoes.

When polarizing generators on cars equipped with double contact regulators disconnect the field lead to prevent damage to the upper regulator contacts.

Circuit "B" generators are polarized by disconnecting the regulator field (F) terminal lead and momentarily touching it to the regulator battery (B) terminal.

The generator field coil resistance will protect the regulator points from even momentary excessive current flow with either method of polarizing. These procedures for polarizing the generator are commonly referred to as "flashing the fields."

Remember — generator polarizing is performed after the regulator leads are connected but BEFORE the engine is started.

polarity ≠ current flowing in right way.

never polarize an alternator!

GENERATOR AND REGULATOR TESTS

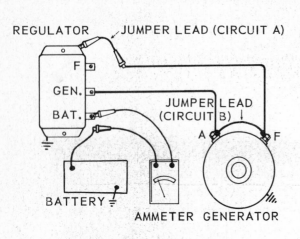

GENERATOR OUTPUT TEST

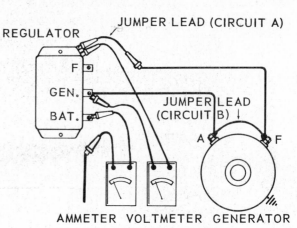

CUTOUT RELAY TEST

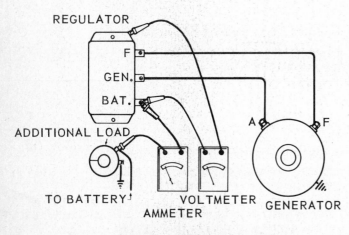

CURRENT REGULATOR TEST

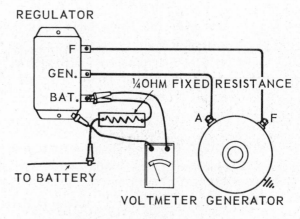

VOLTAGE REGULATOR TEST

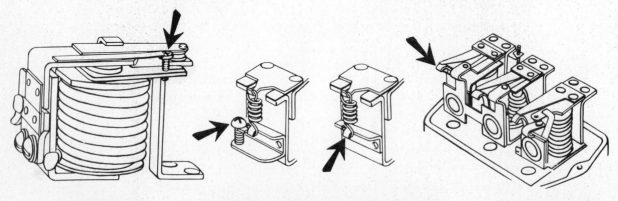

CUTOUT RELAY ADJUSTMENT

ADJUSTING SCREW COIL SPRING

ADJUSTING HANGER COIL SPRING

ADJUSTING ARM FLAT SPRING

REGULATOR ADJUSTMENTS

GENERATOR AND REGULATOR TESTS

The following generator and regulator ON THE CAR test procedures and test equipment hookups are applicable to most charging systems. However, always use the tester hookups and follow the test procedures recommended by the manufacturer of your test equipment.

The charging system should be tested and adjusted only when the units in the system are at operating temperature.

When adjusting the voltage regulator units, make the adjustment very carefully. A slight change in adjustment makes a considerable change in setting.

Test specifications used in this procedure are only averages. Follow the specifications listed for the particular car being serviced.

GENERATOR OUTPUT TEST
1. Check generator drive belt and adjust as required.

2. Circuit "A" only: Disconnect regulator field lead from regulator F terminal and ground this lead to the regulator base with a jumper.

   Circuit "B" only: Clip a jumper lead from the generator output (Arm) terminal to the generator field (F) terminal.

3. Disconnect the regulator battery lead from the BAT terminal and connect a test ammeter between disconnected lead and BAT terminal.

4. Connect tachometer between distributor primary terminal and ground.

5. Turn on all electrical accessories — bright lights, radio, etc.

6. Start engine and slowly increase engine speed until ammeter reads 35 amperes. Engine speed at this time should NOT exceed 1500 rpm. Return engine speed to idle immediately after making test. If generator does not produce rated output at less than 1500 rpm, the generator is defective.

CUTOUT RELAY TEST
1. Leave all test and meter leads connected as in generator output test.

2. Connect voltmeter from regulator Gen (Arm) terminal to ground on regulator base.

3. Start and slow idle engine. Very slowly increase engine speed while observing meters.

   Highest voltage reading just before ammeter indicates charge, is cutout relay closing voltage.

Relay closing voltage is between 12.0 and 13.0 volts. If voltage does not conform to specifications, remove regulator cover, adjust cutout relay, replace cover and retest. Increasing relay spring tension increases the closing voltage. Decreasing the spring tension decreases the closing voltage.

4. Very slowly decrease engine speed while observing ammeter. Lowest ampere reading below zero, just before meter swings to zero, is relay opening counteramperage. Reverse current should be between 1 to 3 amperes.

If reverse current does not conform to specification, the relay air gap and point gap opening must be adjusted to specifications.

CURRENT REGULATOR TEST

1. Circuit "A" only: Reconnect regulator field lead to regulator F terminal.

   Circuit "B" only: Remove jumper lead from generator armature and field terminals.

2. Move voltmeter lead from regulator Gen (Arm) terminal to regulator battery (Bat) terminal. Leave other voltmeter lead grounded to regulator base.

3. Connect a variable load (carbon pile) across the battery. Turn the load control to lowest load (open circuit).

4. Operate engine at 1500 rpm and slowly apply load to battery while observing ammeter for highest reading. Generator output should be between 34 and 38 amperes.

If current regulator setting does not conform to specifications, remove regulator cover, adjust current regulator, replace cover and retest. Increasing the current regulator spring tension increases the current setting. Decreasing the spring tension decreases the current setting.

VOLTAGE REGULATOR TEST

1. Remove ammeter and connect a ¼ ohm fixed resistor between the regulator disconnected battery lead and the regulator battery (Bat) terminal.

2. Operate engine at 2000 rpm and observe voltmeter. Reading should be between 13.5 volts and 14.7 volts.

If voltage regulator setting does not conform to specifications, remove regulator cover, adjust voltage regulator, replace cover and retest. Increasing the voltage regulator spring tension increases the voltage setting. Decreasing the spring tension decreases the voltage setting.

GENERATOR AND REGULATOR
QUICK CHECKS

GENERATOR AND REGULATOR QUICK CHECKS

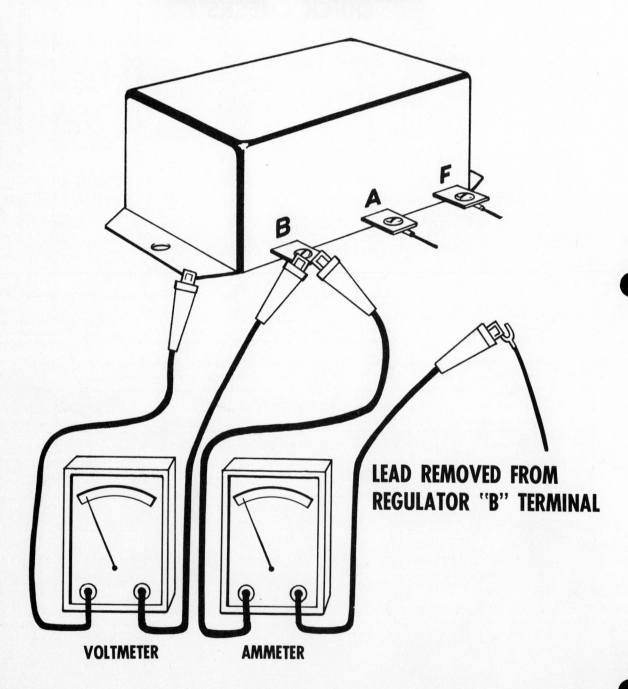

B A F

LEAD REMOVED FROM
REGULATOR "B" TERMINAL

VOLTMETER   AMMETER

GENERATOR AND REGULATOR QUICK CHECKS

Disconnect regulator battery lead and connect ammeter between disconnected lead and regulator "BAT" terminal. Connect voltmeter from regulator "BAT" terminal to ground. Start engine and set engine speed at 1500 rpm.

Note: Observe differences in Circuit "A" and Circuit "B" test procedures where indicated.

METER READING INDICATIONS

- 7.5 volts or more / 15.0 volts or more
  - Regardless of ampere output
    - Remove regulator "F" wire and DO NOT touch to ground
      - Voltage remains high → Repair or replace generator
      - Voltage drops → Check regulator ground. Adjust or replace regulator

- 6.8 to 7.5 volts / 13.8 to 15.0 volts
  - 10 amps. or more
    - Run engine at fast idle for 15 minutes to charge battery
      - 6.8 to 7.5 volts / 13.8 to 15.0 volts / 10 amps. or more → Test battery, generator, regulator and circuitry
      - 6.8 to 7.5 volts / 13.8 to 15.0 volts / 0 to 10 amps. → System OK
  - 0 to 10 amps. → System OK

- 6.8 volts or less / 13.8 volts or less
  - 10 amps. or more
    - Run engine at fast idle for 15 minutes to charge battery
      - 6.8 to 7.5 volts / 13.8 to 15.0 volts / 0 to 10 amps. → System OK
      - 6.8 volts or less / 13.8 volts or less / 0 to 10 amps. → Check regulator ground. Adjust or replace regulator
      - 6.8 volts or less / 13.8 volts or less / 10 amps. or more → Test and/or charge battery
  - 0 to 10 amps.
    - Circuit "A": Remove regulator "F" wire and ground it. Circuit "B": Place jumper lead from "A" to "F" terminals on generator
      - Voltage and current increase → Check regulator ground. Adjust or replace regulator
      - Voltage remains low → Repair or replace generator

CHARGING SYSTEM RESISTANCE TESTS

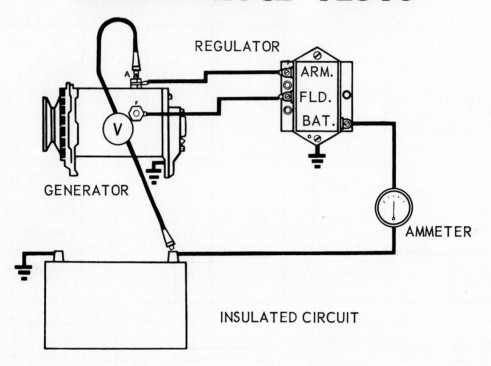

INSULATED CIRCUIT

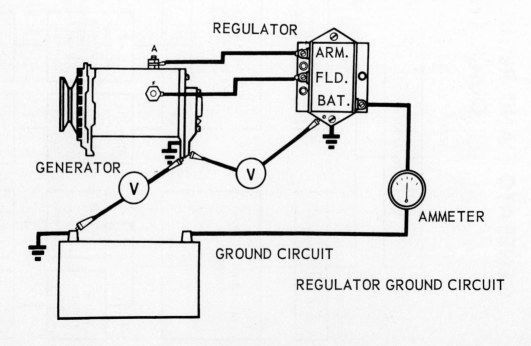

GROUND CIRCUIT

REGULATOR GROUND CIRCUIT

CHARGING SYSTEM RESISTANCE TESTS

The charging system is designed for minimum voltage loss due to resistance in its circuitry. Excessive resistance in any part of the charging system circuit will alter its efficiency proportionately.

The insulated circuit test is made to determine the amount of voltage loss occurring between the armature terminal of the generator and the insulated battery post. The ground circuit test is made to determine the amount of voltage loss between the battery ground post and the generator housing. The regulator ground circuit test is made to determine the amount of voltage loss occurring between the regulator base and the generator housing.

Any voltage loss caused by high resistance in these circuits will reduce the overall charge rate and can lead to eventual battery discharge. High resistance can be present in the form of poor connections or defective wiring. If excessive resistance is indicated by a test, it should be located and corrected.

The maximum permissible voltage loss in the charging system insulated circuit, with a 20 ampere charge flowing, is 1.0 volt. If more than 1.0 volt is indicated, the cause of the excessive resistance should be isolated and corrected permitting only .2 volt for each wire and each regulator terminal in the circuit.

The maximum permissible voltage loss in the charging system ground circuit, with a 20 ampere charge rate flowing, is .2 volt. Higher voltage drop readings indicate a poor ground connection between the battery and the engine or between the generator and the engine.

The permissible voltage drop in the regulator ground circuit, with a 20 ampere charge rate, is .2 volt. Excessive voltage loss due to high resistance will cause the cutout relay and voltage regulator to operate at higher settings than those for which they are actually adjusted. This test is particularly critical because the regulator contact points, the voltage windings on the relays and the resistance units are all grounded in the regulator. A perfect ground circuit for these units is **essential** to proper regulator operation.

ALTERNATOR CHARGING SYSTEM

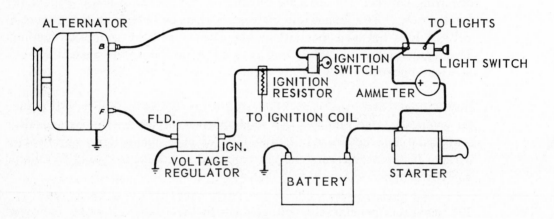

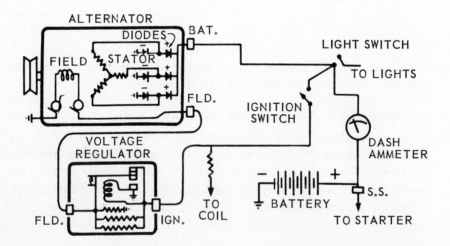

ALTERNATOR CHARGING SYSTEM

The alternator charging system has been developed to answer the need for increased generator output necessitated by the addition of electrically-operated accessories on the modern car combined with low-speed driving in congested traffic.

The alternator possesses the ability to produce a current output at engine idle speed and at low car speed. This factor makes it a superior charging unit to the direct current generator which must be rotated at reasonable speed before a current output is developed.

The basic alternator charging system components include the battery, the self-rectifying alternating current generator, a voltage limiting relay and interconnecting wiring. The circuit will include either an ammeter or an indicator light.

Every AC charging system is controlled by a voltage regulator. The battery initially supplies the current for the alternator field coil. At this time, and during idle and low-speed operation, there is no voltage control problem. However, as alternator speed increases with engine speed, the voltage increase developed by the alternator would be imposed on the field coil. This increases the field strength and further raises the voltage output. Unless this voltage rise condition is kept under control, the high voltage developed will result in damage or shortened life expectancy of light bulbs, relay coils, radio tubes, turn signal flashers, breaker points and other voltage-sensitive units.

All alternator voltage regulators are of the double-contact type discussed under generator regulators.

In other alternator systems the regulator may contain, in addition to the voltage limiter, a field relay which is used to complete the field circuit when the ignition switch is turned on. It may also include a lamp relay when an indicator light is used instead of an ammeter.

The cutout relay used in the DC generator system is not used in AC systems because the diode rectifiers used in the AC system permit the flow of current in one direction only. The positive diodes will not allow current to flow from the battery into the alternator. Battery discharge is thereby prevented.

The AC system does not require the use of a current regulator since the alternator is self-limiting in current output as long as voltage control is maintained.

The chart illustrates a schematic arrangement of an AC charging system in common use. The field terminal of the alternator is connected to the field terminal of the regulator. The connection from the ignition terminal of the regulator goes to the ignition switch. Field current must be supplied from the battery, as the rotor of the alternator does not possess residual magnetism. The lead from the output terminal completes the circuit from the alternator to the battery positive terminal.

The car ammeter is in the circuit between the ignition switch and the battery. In that location it will register charge only when the alternator output is greater than the electrical load. The ammeter will register discharge only when the alternator output is less than the electrical load. The ammeter does not register the alternator output. It registers only the current flow into or out of the battery.

The AC charging system illustrated on this chart essentially covers the elements of all AC systems. Individual coverage of specific systems will be covered under testing procedures a little later in this course.

ALTERNATOR COMPONENTS

p. 145

- when the voltage regulator lower contacts are closed, the rotor is supplied with an unrestricted flow of current, resulting in maximum output.

- when ~~both~~ both the upper and lower contacts are open, current to the rotor is supplied through a 14-ohm ballast resistor, resulting in reduced output.

- when the upper contacts are closed, there is no current supplied to the rotor circuit, resulting in NO output.

ALTERNATOR COMPONENTS

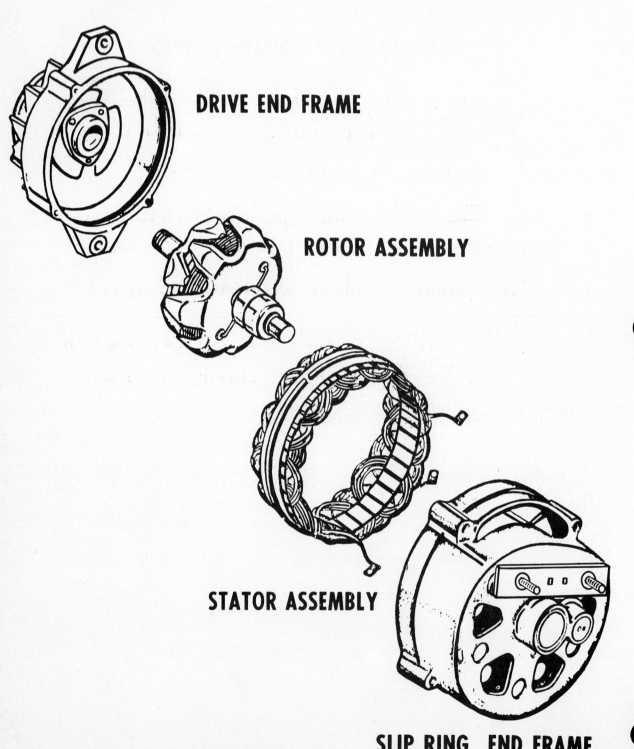

DRIVE END FRAME

ROTOR ASSEMBLY

STATOR ASSEMBLY

SLIP RING END FRAME

ALTERNATOR COMPONENTS

The alternator is composed of a rotor assembly, a stator assembly and two end frame assemblies, one at the drive end of the alternator and one at the slip ring end.

The rotor assembly is composed of a field coil made of many turns of wire wound over an iron core which is contained between two iron segments with interlacing fingers. These fingers serve as magnetic poles. This assembly is press mounted on a steel shaft which turns in prelubricated anti-friction bearings. Two slip rings are mounted on one end of the shaft.

Each end of the field coil winding is connected to one of the slip rings. A brush rides on each slip ring. These brushes conduct battery current to the rotor winding to create the magnetic field required for current generation. This is necessary because the rotor is made of a metal alloy that does not retain residual magnetism and is therefore not self-exciting.

The stator assembly is composed of a laminated iron frame and three sets of windings wound into slots in the frame. The manner in which these windings are wound and connected makes the alternator a three phase unit. Constant voltage can be maintained since each of the three windings reaches its maximum voltage output at a different time. Three voltage impulses are induced for every turn of the rotor with this arrangement whereas one voltage impulse would be induced if a single winding were used. One end of all three windings are connected together while the other end of each winding is connected to a pair of diodes, one positive and one negative.

When assembled, a very small air gap is present between the rotor poles and the stator to keep the magnetic field lines of force as strong as possible. As the rotor spins, the alternate north and south poles of the rotor fingers pass each loop in the stator windings inducing an alternate amperage and voltage in the windings. This alternating current is then rectified by the diodes.

The slip ring end frame contains six diodes, which are electrical rectifying devices. The diodes, three negative and three positive, act as one-way valves permitting current to pass freely in one direction but not in the other. By their combined action, the alternating current generated is rectified to direct current.

DIODE

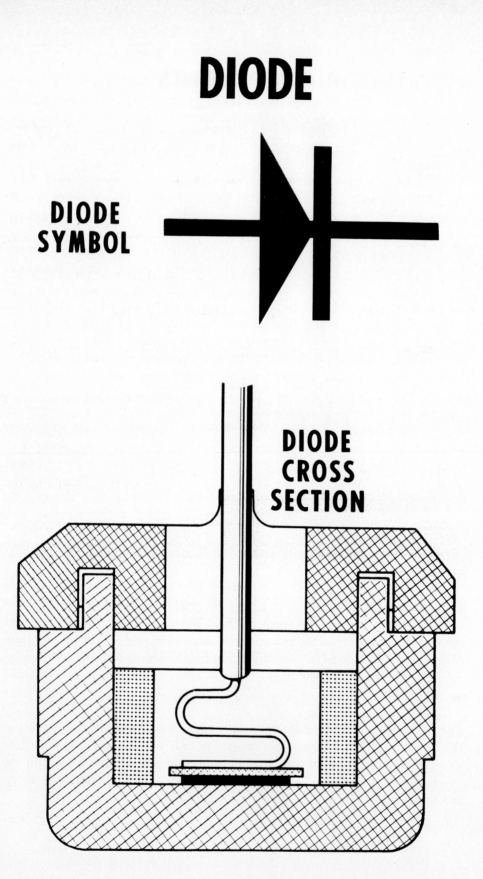

**DIODE
SYMBOL**

**DIODE
CROSS
SECTION**

DIODE

As previously stated, the diode is a current rectifying device. It serves as a one-way electrical check valve which permits current to flow readily in one direction but stops its flow in a reverse direction. The silicon die or wafer in the diode possesses this electrical characteristic by virtue of the molecular construction of the metal.

The diode symbol is an arrow indicating the direction of current flow allowed by the diode. The bar indicates a one-way "gate" or block to current flowing in the opposite direction.

The cross-sectional view illustrates the position of the silicon wafer in the bottom of the diode case. The case is made of rather heavy metal to serve both as a protection for the rather brittle silicon wafer and to effectively dissipate the heat, induced by the current flow through the diode. The case is tightly sealed during manufacture to prevent the entrance of moisture into the diode which would result in short circuiting of the unit. Moisture is readily drawn into any unit that operates at a temperature since it "breathes" as it heats and cools.

In all negative ground alternator charging systems the negative (case) diodes are pressed into the alternator grounded end frame and the positive (case) diodes are pressed into a holder called a heat sink. The heat sink is usually made of die-cast aluminum because it possesses high heat dissipating qualities. It is mounted in, but electrically insulated from, the end frame. The end frame also serves to absorb the heat developed by the passage of current through the diodes.

A negative diode is properly identified as a negative case diode meaning the diode case is negative polarity and the diode lead is positive polarity. A positive case diode will have a positive polarity case and a negative polarity lead. Diodes are color coded for polarity identification. The part number of the diode is printed with a red dye on the positive diode and with a black dye for the negative diode. When part numbers are not used, a dab of red or black dye on the diode case identifies its polarity. Diodes used in heavy-duty application are identified with a + (plus) or a – (minus) sign.

The Zener diode is used in many transistorized ignition systems and voltage regulators. Its basic function is to protect the transistors in the circuit from the harmful effects of high voltage. When a predetermined voltage is reached, the Zener diode "breaks down" and permits the passage of current in the opposite direction by providing a shunt circuit for the high-voltage current. This breakdown voltage does not harm the Zener diode since it is designed to perform in this manner. When the voltage drops below the predetermined voltage, the Zener diode again blocks current.

Since the breakdown voltage of the Zener diode is lower than the voltage value that would damage the transistor, the transistor is thereby protected. The Zener diode can be constructed to "break down" at varying voltages.

TRANSISTORS

APPEARANCE

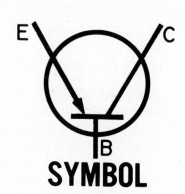

SYMBOL

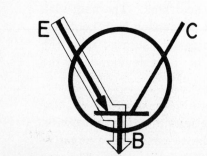

EMITTER-BASE CIRCUIT

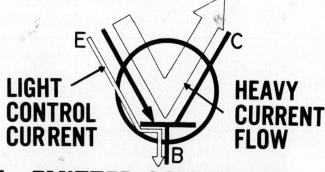

LIGHT CONTROL CURRENT

HEAVY CURRENT FLOW

EMITTER-COLLECTOR CIRCUIT

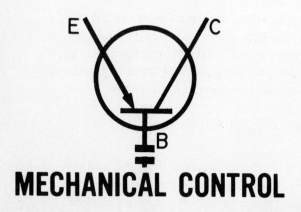

MECHANICAL CONTROL

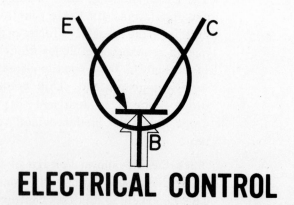

ELECTRICAL CONTROL

TRANSISTORS

A transistor is a device which acts as an electrical switch but has no moving parts. Current flow through the transistor can be controlled mechanically by a set of contacts or electrically by reversing the circuit polarity.

A transistor is made up of three small sections of material fused together like a wafer or sandwich and placed in a container. The sections are referred to as ''N'' or ''P'' type material. The arrangement of the sections determines the transistor's polarity. The transistor may be known as a PNP transistor or as a NPN transistor according to the positions of the wafers. The middle wafer serves as the base and therefore dictates the polarity.

The transistor symbol is illustrated on the chart. The circle represents the container. The letters represent the three elements of which the transistor is composed — E is the Emitter; C is the Collector; and B is the Base. The arrow on the Emitter symbol indicates the direction of current flow through the transistor.

There are two important factors relative to the manner in which the transistor works. First: the Emitter-Collector circuit is the main current-carrying circuit. Second: current flow through the Emitter-Collector circuit is possibly ONLY when there is current flow through the Emitter-Base circuit. Although the current flow in the Base circuit may be only a fraction of the current flow in the Collector circuit, the Collector circuit cannot exist without the Base circuit. It follows then, that an interruption of the light current flow in the Base circuit will cause a stoppage in the heavy current flow in the Collector circuit. In this manner the Base circuit ''triggers'' the transistor and turns it ON or OFF.

As previously stated, the transistor is electrically controlled through the base circuit. A set of mechanical contact points, in series in the Emitter-Base circuit, can readily carry the light current flow in this circuit and serve to ''make'' and ''break'' the Base circuit.

Another method of electrical control is performed by reversing the current flow in the Base circuit. The Base circuit will carry current only when the proper voltage range and polarity is applied. When incorrect voltage or reversed polarity is applied the Base circuit is turned OFF and current flow in the Collector circuit is consequently stopped.

DIODE-RECTIFIED
THREE PHASE OUTPUT

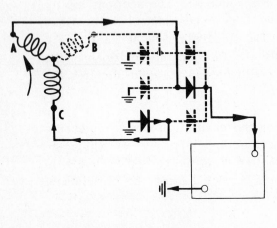

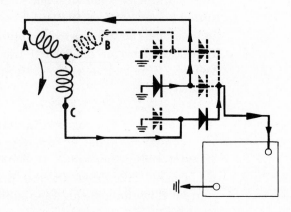

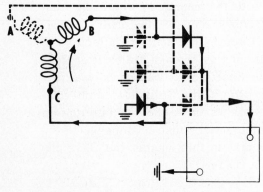

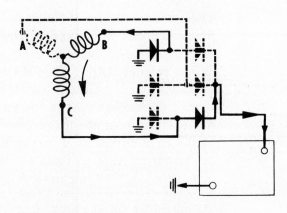

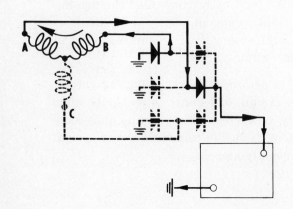

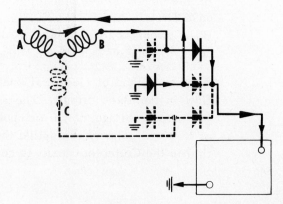

DIODE-RECTIFIED OUTPUT

This chart illustrates how the diodes rectify the alternating current flowing in either direction through any two stator windings so that direct current is always available at the output or battery terminal of the alternator.

Voltage induced in the windings "A" and "C" cause current to flow from the "A" coil of the stator through its positive diode, out of the "Bat" terminal of the alternator and to the battery. The return circuit is from the battery, to the alternator ground terminal, through the negative diode connected to the "C" coil of the stator and back to the "A" coil where it started, completing the circuit.

When the alternating voltage induced in these windings is reversed, current will flow out of the "C" stator coil of the alternator, through its positive diode, out of the "Bat" terminal of the alternator and to the battery. This rectified direct current flows to the battery in the same direction as it did previously. The return circuit is from the battery, to the alternator ground terminal, through the negative diode connected to the "A" stator coil and back to the "C" coil where it started, completing the circuit.

Current flow, in either direction, through the stator windings in the other illustrations can be traced by following the arrows from the stator coil windings, through the positive diode, to the battery, from the battery through the negative diode, and through the stator windings back to its source.

These illustrations clearly show that by the action of the diodes, direct current is always flowing into the battery regardless of which stator windings are inducing the voltage or the direction of the current flow through the windings.

Occasionally a diode is used in an alternator for a function other than current rectification. An example is the isolation diode used in the Motorola alternator. Its function is to serve as a load relay to essentially disconnect the alternator from the battery when the ignition switch is turned off. In a sense it is serving the same function as a cutout relay in a DC charging system. When the vehicle is not in use, this diode eliminates the possibility of electrical leakage over the insulators.

AC CHARGING SYSTEM
With Ammeter

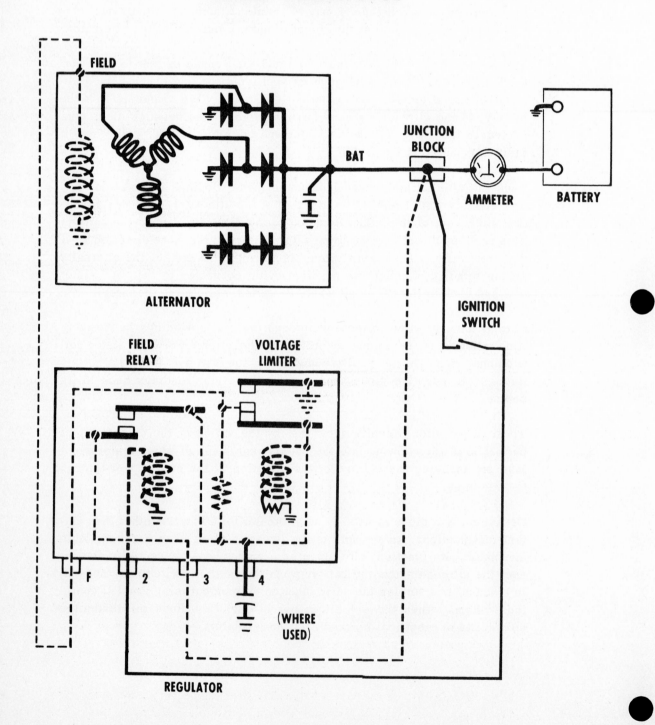

FIELD

ALTERNATOR

BAT

JUNCTION
BLOCK

AMMETER

BATTERY

IGNITION
SWITCH

FIELD
RELAY

VOLTAGE
LIMITER

F

2

3

4

(WHERE
USED)

REGULATOR

AC CHARGING SYSTEM (With Ammeter)

A typical AC charging system equipped with an ammeter is illustrated on this chart.

When the ignition switch is closed, a connection is established between its battery and ignition terminals. Battery current now flows to terminal "2" of the regulator, through the field relay shunt winding, to ground and back to the battery through the ground circuit.

The magnetic field created around the field relay core attracts the relay armature thereby closing the relay points. Battery current now flows to terminal "3" of the regulator, across the field relay points, across the lower voltage regulator points, from the field terminal on the regulator to the field terminal on the alternator, through the field coil to ground and back to the battery. Some current also flows through the shunt winding on the voltage coil. With the rotor field coil energized, the alternator is ready to produce current as soon as the rotor is turned.

With the field relay closed, current is supplied directly to the alternator field coil from the battery instead of through the ignition switch and ignition primary resistance wire.

When the engine is started, the alternator rotor spins and the magnetism created in the field coil by the field current induces an alternating current in the stator windings. The diodes rectify the alternating current generated into direct current as previously explained.

As the vehicle is put in motion, the alternator speed is increased resulting in a greater voltage being induced on the current flow in the field circuit. Since the shunt winding on the voltage coil is also subjected to this increased voltage, a greater magnetic field is created around the voltage coil. This strong field attracts the voltage regulator armature causing the lower contact points to separate.

Field current must now flow through the resistor on its way to the field coil. Field current is reduced from approximately 2 amperes to about ¾ ampere by the resistor. The reduced field current results in an immediate reduction in alternator output with an associate drop in voltage applied to the voltage regulator coil. The voltage regulator armature spring closes the lower contacts thereby re-establishing full field current flow. This cycling action of inserting and removing the resistor from the field circuit limits the voltage developed by the alternator to a safe value.

If the vehicle is driven at high speed and the accessory and battery demands are low, a higher voltage of .1 to .3 volt is induced on the shunt coil of the voltage limiter. This results in the upper armature being attracted to the relay core thereby closing the upper contacts. At this time, both ends of the field coil are grounded with the result that there is no current flow through the coil. With a "dead" field coil, alternator voltage decreases permitting the upper contact points to open. Field current now flows through the resistor to the field coil. As the voltage again increases, the upper relay contacts are again closed. The cycling that takes place limits the field current between ¾ ampere and no current at all. By this action, alternator output is safely limited regardless of how fast the vehicle may be driven or how long the speed is sustained.

The function of the condenser used in the alternator is to dampen the high voltage surges developed in the stator windings or any transient high voltage impulses generated anywhere in the charging system. The condenser also serves as a noise suppression unit.

AC CHARGING SYSTEM
(With Indicator Lamp)

AC CHARGING SYSTEM
With Indicator Lamp

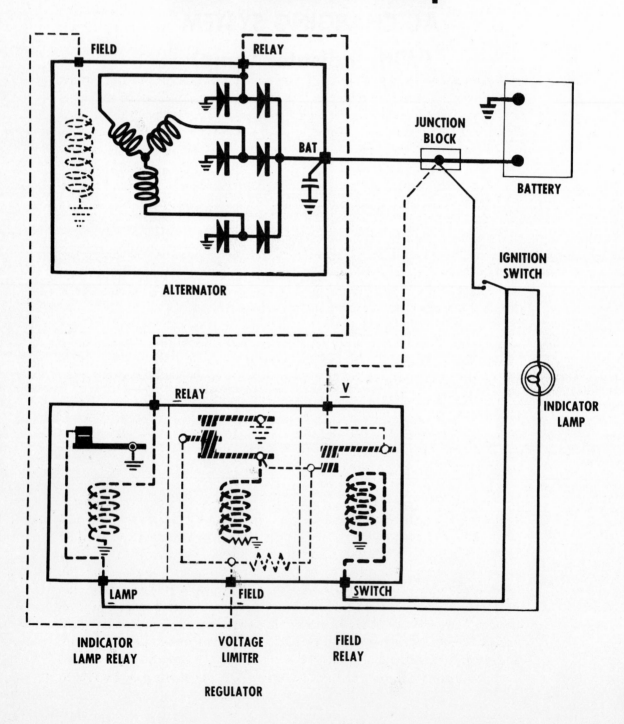

FIELD

RELAY

BAT

JUNCTION
BLOCK

BATTERY

ALTERNATOR

IGNITION
SWITCH

RELAY

V

INDICATOR
LAMP

LAMP

FIELD

SWITCH

INDICATOR
LAMP RELAY

VOLTAGE
LIMITER

FIELD
RELAY

REGULATOR

AC CHARGING SYSTEM (With Indicator Lamp).

A typical AC charging system equipped with an indicator lamp is illustrated on this chart.

When the ignition switch is closed, battery current flows through the indicator lamp to the "L" terminal on the regulator, across the closed relay contacts and to ground. This completes the indicator lamp relay circuit permitting the lamp to light.

Also energized by the closing of the ignition switch is the field relay. Battery current flows from the switch to the "SW" terminal on the regulator, through the field relay voltage coil, to ground and back to the battery. The magnetic field created around the field relay core attracts the relay armature closing the relay points. Battery current now flows from the junction block terminal to the "V" terminal on the regulator, across the field relay points, across the lower voltage regulator points, from the "F" terminal on the regulator to the alternator, through the field coil to ground and back to the battery.

When the engine is started, the alternator rotor spins and the magnetism created in the field coil by the field current induces an alternating current in the stator windings. The diodes rectify the alternating current generated into direct current as previously explained.

As soon as the alternator starts operating, current flows from the alternator "relay" terminal to the "R" terminal on the regulator, through the voltage coil on the indicator lamp relay to ground and back to the alternator. This current flow magnetizes the light relay core which attracts the relay armature thereby opening the relay points. This opens the indicator lamp circuit, and the lamp goes out. This circuit arrangement provides a light which serves as a warning when lit with the engine running, that trouble exists in the charging system.

As the vehicle is put in motion, the alternator speed is increased resulting in a greater voltage being induced on the current flow in the field circuit. Since the shunt winding on the voltage coil is also subjected to this increased voltage, a greater magnetic field is created around the voltage coil. This strong field attracts the voltage regulator armature causing the lower contact points to separate. Field current must now flow through the resistor on its way to the field coil. Field current is reduced from approximately 2 amperes to about ¾ ampere by the resistor. The reduced field current results in an immediate reduction in alternator output with an associate drop in voltage applied to the voltage regulator coil. The voltage regulator armature spring closes the lower contacts thereby re-establishing full field current flow. This cycling action of inserting and removing the resistor from the field circuit limits the voltage developed by the alternator to a safe value.

If the vehicle is driven at high speed and the accessory and battery demands are low, a higher voltage of .1 to .3 volt is induced on the shunt coil of the voltage limiter. This results in the upper armature being attracted to the relay core thereby closing the upper contacts. At this time, both ends of the field coil are grounded with the result that there is no current flow through the coil. With a "dead" field coil, alternator voltage decreases permitting the upper contact points to open. Field current now flows through the resistor to the field coil. As the voltage again increases, the upper relay contacts are again closed. The cycling that takes place limits the field current between ¾ ampere and no current flow at all. By this action, alternator output is safely limited regardless of how fast the vehicle may be driven or how long the speed is sustained.

AC CHARGING CIRCUIT
INDICATOR LAMPS

AC CHARGING CIRCUIT INDICATOR LAMPS
Indicator lamp with separate lamp relay

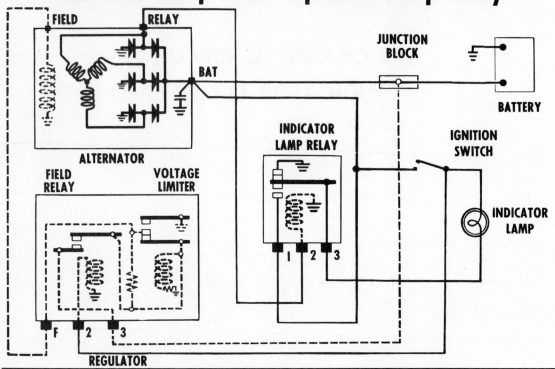

Indicator lamp without lamp relay

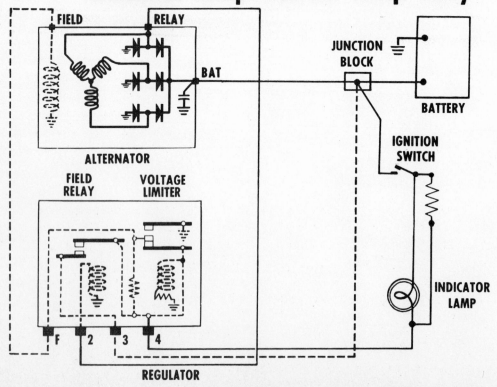

AC CHARGING CIRCUIT INDICATOR LAMPS

The upper circuit diagram illustrates the use of an indicator lamp and a separate indicator lamp relay.

When the ignition switch is turned on, battery current flows through the indicator lamp, to the indicator lamp relay No. 3 terminal, across the relay armature and the closed upper relay points to ground. This completes the circuit permitting the indicator lamp to light.

As soon as the engine starts, alternator voltage from the relay terminal is impressed on the indicator lamp relay No. 2 terminal. The current flows through the relay winding creating a magnetic field which attracts the relay armature pulling it down and closing the lower contacts.

When this occurs, not only is the circuit ground denied to the indicator lamp but alternator output voltage is now applied to both sides of the lamp at the same time. As a result, current flow stops and the light goes out.

The rest of the charging system functions are as previously explained.

The lower circuit diagram illustrates the use of an indicator lamp without the use of an indicator lamp relay.

When the ignition switch is turned on, battery current flows through the indicator lamp to the regulator No. 4 terminal, across the closed lower voltage regulator points, to the regulator "F" terminal and to the alternator field coil. This complete circuit permits the indicator lamp to light.

As soon as the engine starts, alternator voltage is impressed on the alternator relay terminal causing current to flow to the regulator No. 2 terminal and through the shunt winding in the field relay. The magnetic field created by this current flow attracts the field relay armature closing the field relay circuit. System voltage impressed on the regulator No. 3 terminal is now also impressed on the No. 4 terminal. This results in system voltage being applied to both sides of the indicator lamp at the same time. With no current flow through the lamp, the light goes out.

In this regulator circuit, it may be said the field relay has a dual function. It not only completes the field circuit directly from the battery to the alternator field instead of through the ignition switch and primary resistance wire but it also serves as an indicator light relay.

The rest of the charging system functions as previously explained.

ALTERNATOR TESTING FACTORS

Failure of the AC charging system to function normally is revealed by an indicator lamp that does not light when the ignition switch is turned on; by a lamp that stays lit after the engine starts running; by a slower than normal cranking speed; or by a battery being in a state of undercharge or overcharge.

An alternator with a faulty diode can put out enough current to supply the ignition system demands and yet be incapable of keeping the battery fully charged especially when the lights and accessories are used.

When an alternator diode is defective, an obvious indication is a whine or hum with the engine idling or operating at low speed. Since the alternator is a 3-phase machine, when one diode is defective the machine is out-of-phase, soundwise, resulting in a whine. This condition is usually the result of a shorted diode. When a diode is open, the condition is generally indicated by noisy operation of the alternator. This noise is caused by the physical unbalance of the unit which has been created by the electrical unbalance, so to speak. If this condition is allowed to persist, the antifriction rotor shaft bearings in the diode end frame may be damaged.

By far, the greatest percentage of alternator trouble is diode trouble. But usually this trouble is created by improper test procedures, reverse current connections, removing alternator leads while the alternator is in operation, reversing battery connections and other abuses.

When an AC charging system complaint is expressed, a few checks and tests should be made before the alternator is condemned or disassembled.

1. Check the tension of the drive belt and inspect its condition. Stretched, frayed or oil-wetted belts should be replaced. The smaller alternator drive pulley has less wraparound drive belt action making belt tension particularly critical. Further, since the alternator has an output even at idle, it is possible for the belt to be slipping at idle speed. Care must be exercised when adjusting belt tension so that the aluminum alternator housing is not crushed by the pry bar. Rest the bar only on the heavy front section of the housing.

2. Test the condition of the battery. A sulphated or internally defective battery will resist being charged even when the charging system is functioning normally.

3. Excessive resistance in the charging circuit can cause a lower than normal charge rate and result in a discharged battery. To isolate the point of high resistance with your test equipment, test both the insulated circuit and the ground circuit.

4. Make an alternator field current draw test and an output test.

5. Test the regulator. A malfunctioning field relay may be restricting field current thereby reducing alternator output. A low voltage regulator setting can also be responsible for an undercharged battery condition.

If the above checks and tests do not reveal any defective conditions, complete tests of the alternator should be conducted before the unit is disassembled.

Starting with the 1972 models, be sure the heater/air condition blower motor is disconnected, along with all the other electrical accessories, before conducting an alternator output test. Pulling the air conditioner fuse is a quick way to cut the power to the blower motor.

To avoid the hazzard of instant, accidental fogging of the windshield, the blower motor operates continuously at low speed unless regulated otherwise. Failure to disconnect the blower motor before testing will result in a lower then specified output since as much as 10 amperes can be flowing in the blower circuit. This reduced ammeter reading may be misinterpreted as a lack of sufficient output even though the charging system is functioning properly. The discrepancy occurs because part of the alternator output is going through the blower motor circuit without passing through the test ammeter since the two circuits are in parallel.

TRANSISTOR REGULATORS

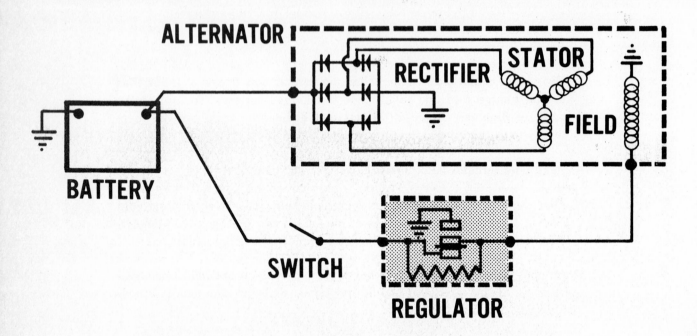

VIBRATING POINT REGULATOR

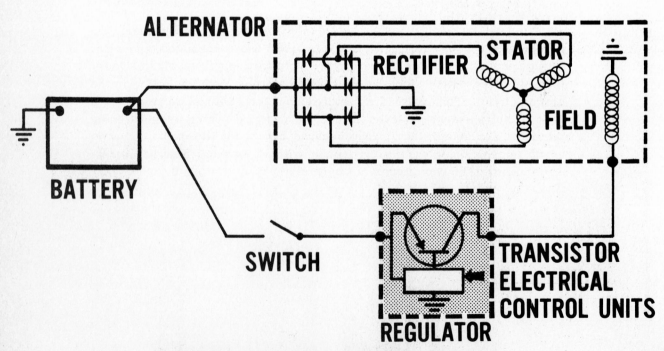

TRANSISTOR REGULATOR

TRANSISTOR REGULATORS

The transistorized voltage regulator controls the alternator voltage output electronically by using transistors, diodes, resistors and a capacitor. Some transistorized regulators employ a vibrating contact field relay to control the transistor which carries the field current. A vibrating voltage limiter relay may also be used in conjunction with a transistor. When the regulator design eliminates the use of relays and employs full transistor control, the regulator is called a "solid state" regulator.

The chart illustrates the elements of a simplified alternator charging system. In the regulator control circuit is housed the thermistor, a temperature-compensating voltage control device; a Zener diode, a voltage sensing unit; a driver transistor, which controls the output transistor; resistors and a capacitor. All these units function together to control the alternator field current and consequently alternator output.

When the ignition switch is closed, battery voltage is applied to the emitter-collector circuit of the transistor and to the alternator field circuit. The emitter-collector circuit is completed because the emitter-base circuit is also completed through the regulator control circuit. Alternator voltage builds up as soon as the engine starts, supplying all the activated circuits with current and charging the battery. As the vehicle is put in motion, alternator voltage builds up to the value for which the regulator has been set. The regulator control circuit then applies a higher voltage (or reverse voltage) to the base circuit of the transistor thereby turning the transistor "Off." The resultant loss of current flow in the emitter-collector circuit causes a stoppage of current flow in the alternator field circuit with a resultant drop in alternator output.

The regulator control circuit now places a lower voltage on the base circuit of the transistor allowing the transistor to be turned "On" again. Current flow in the emitter-collector circuit and in the alternator field circuit is reestablished and alternator output voltage again builds up to the value of the voltage regulator setting. The regulator control circuit again reverses the voltage applied to the transistor base circuit turning the transistor "Off." In this manner, the transistor is switched "On" and "Off" regulating the alternator output to match the electrical demands of of the vehicle and the charging requirements of the battery.

The vibrating contacts of the mechanical voltage limiter relay vibrate from 75 to 250 cycles per second. The "On" and "Off" cycles of the transistor may occur as frequently as 7,000 cycles per second.

Some transistor regulators are equipped with an external method of adjustment while others are nonadjustable.

ALTERNATOR AND REGULATOR TESTS

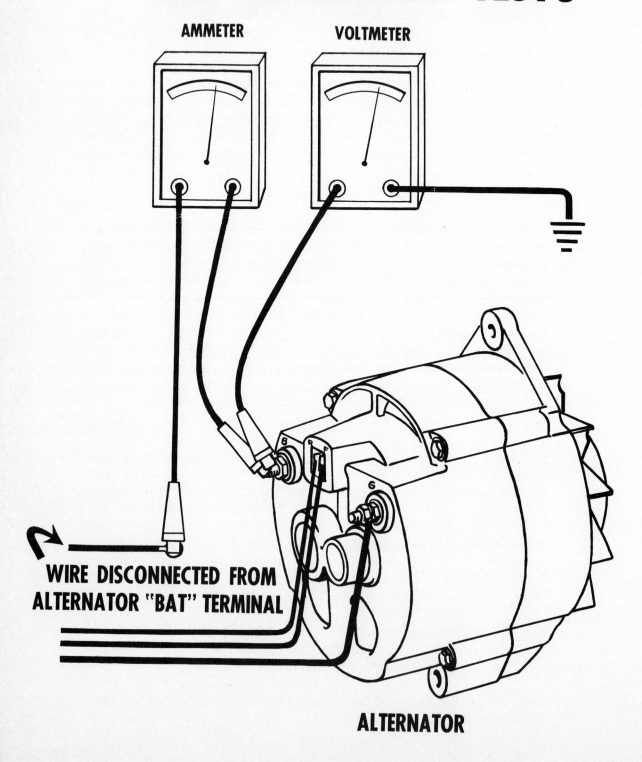

AMMETER

VOLTMETER

WIRE DISCONNECTED FROM
ALTERNATOR "BAT" TERMINAL

ALTERNATOR

ALTERNATOR AND REGULATOR TESTS

In the event different test procedures are suggested, it is advisable to follow the instrument hook-up and test procedure recommended by the manufacturer of your test equipment.

1. Check alternator drive belt and adjust as required.

2. Remove lead from alternator output (Bat) terminal and connect ammeter negative lead to disconnected alternator lead and ammeter positive lead to alternator output (Bat) terminal.

3. Connect voltmeter positive lead to alternator output (Bat) terminal and voltmeter negative lead to ground.

4. Connect tachometer between distributor primary terminal and ground.

5. Start and operate engine at 1500 rpm.

6. Observe meters. Voltmeter should read between 13.5 volts and 15 volts. Ammeter should read approximately 10 amperes.

If the voltmeter reading is less than 13 volts, disconnect the field lead from the alternator being careful not to touch it to ground. Connect a jumper from the alternator field terminal to the alternator output (Bat) terminal. If a higher voltage and amperage is obtained, a defective regulator is indicated. If voltage remains low, the alternator is defective.

If voltmeter reading is more than 15 volts, disconnect the field lead from the alternator being careful not to touch it to ground. If voltmeter reads 12 volts (battery voltage) and the ammeter reads zero amperes, the regulator is defective.

The alternator charging system circuitry is tested similarly to the DC charging system for insulated, ground and regulator ground resistance. In addition, a field circuit resistance test is also conducted. When testing alternators, regulators and charging system circuitry, always follow the meter hook-up and test procedure recommended by the manufacturer of your test equipment.

Regulator Adjustment Precautions:

When removing or replacing the regulator cover, the ignition switch must be turned off to avoid accidental shorting.

Regulator settings must be made with an insulated bending tool. Make regulator adjustments very carefully. A slight change in adjustment makes a considerable change in setting.

Final voltage limiter settings must be checked with the regulator cover in place.

ALTERNATOR AND REGULATOR QUICK CHECKS

Disconnect alternator output "BAT" lead and connect ammeter between disconnected lead and alternator "BAT" terminal. Connect voltmeter to alternator output "BAT" terminal and ground. Start engine and set engine speed at 1500 rpm.

Caution: DO NOT allow voltage to exceed 16 volts. If voltage appears as though it would easily exceed 16 volts, check ammeter connections.

Alternator MUST NOT be operated without a load connected to the output terminal.

METER READING INDICATIONS

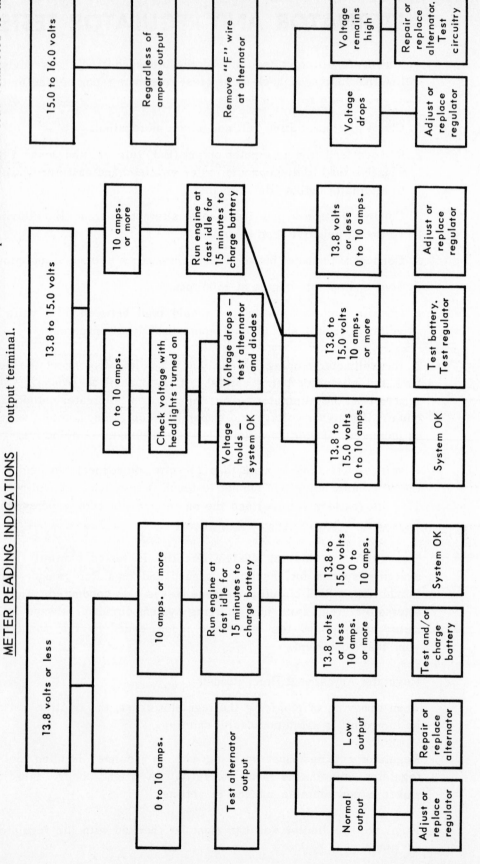

MICRO-CIRCUIT VOLTAGE REGULATORS

MICRO-CIRCUIT VOLTAGE REGULATORS

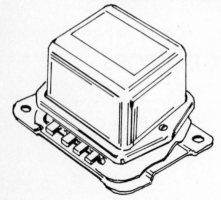

CONVENTIONAL REGULATOR

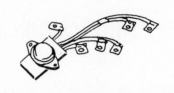

GENERAL MOTORS UNIT

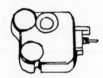

FORD MOTORS UNIT

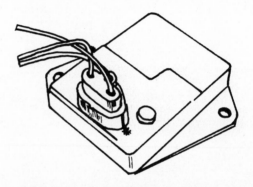

**CHRYSLER CORPORATION
(FIREWALL MOUNTED)**

INTEGRATED-CIRCUIT REGULATOR

SHIELDED BRUSH ASSEMBLY

PROTECTED TERMINALS

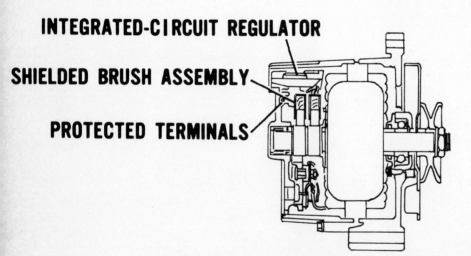

**INTEGRATED MICRO-CIRCUIT REGULATOR
(MOUNTED IN ALTERNATOR HOUSING)**

Chart No. 72

MICRO-CIRCUIT VOLTAGE REGULATORS

The recently introduced miniature solid-state micro-circuit voltage regulators make possible combining some regulators with the alternator into one compact unit. With the elimination of the conventional regulator unit the standard wiring harness between the two units has also been eliminated.

Although construction and design varies, the micro solid-state regulator is basically composed of transistors, diodes and resistors all fabricated within a single piece of silicon crystal measuring about 1/8 inch square. The parts are interconnected by means of very small aluminum conductors and the entire assembly may be fully encased in high-temperature thermo-setting plastic for trouble-free, long-life service. Terminals built into the regulator housing, supply the alternator circuit connecting points. Being factory precision adjusted, the miniature voltage limiter does not need, nor is any provision made for, periodic adjustment. Failure to conform to test specifications, calls for unit replacement.

Several benefits have been attributed to the new regulator design.

1. Advanced technology has miniaturized the voltage regulator for assembly on or inside the alternator. The alternator housing has been finned to provide adequate cooling for both the alternator and the regulator.

2. Simplified circuitry reduces the number of individual components in the charging system.

3. Since the conventional wiring harness between the alternator and regulator has been dispensed with, all wiring and connections are contained within the alternator housing. This arrangement reduces the possibility of voltage loss in the charging system due to loose or poor wiring connection.

4. The regulator solid-state circuitry provides improved voltage control because moving parts as the conventional vibrating contacts have been eliminated. The voltage setting now remains precise for the life of the unit without maintenance or adjustment.

5. The solid-state control sensors are unaffected by shock, vibration, moisture or aging and the long life and inherent accuracy of the semi-conductor elements make this type of circuitry particularly adaptable to the automotive charging system.

6. By maintaining positive and precise control over the alternator output at all temperatures, the battery will be kept charged at all times and increased lamp life will also be realized.

The micro-circuit transistorized Electronic voltage regulator introduced on some 1969 Chrysler Corporation vehicles is mounted on the firewall. If the unit is replaced it is important that the same screws that retained the original regulator be used to secure the new unit. The screws are designed for a special "biting" action to assure a definite electrical ground to the fire-wall. The charging system will fail to function if an effective ground is not secured at the regulator.

Consistent with the policy of regulator nonadjustment is the conventional charging system regulators used on many models of Ford Motor Company products starting with the 1968 models. Although not of micro-circuit design the regulators have the cover riveted to the regulator base. To further identify these regulators as nonadjustable, the covers are painted a blue color. Any regulator that does not test to a specified tolerance should be replaced, not serviced.

Starting with the 1969 Ford products, some models were equipped with the micro-electronic integral regulator which is externally mounted on the alternator rear end housing. Also starting with some 1969 models, General Motors introduced the Delcotron CSI (Charging System Integral) in which the miniature integrated regulator is internally housed in the rear upper section of the alternator.

This policy of nonadjustment is predicated on the fact that regulator tailored voltage adjustments that were necessary on dc charging systems are not required on ac charging systems. It was therefore decided that a factory-adjusted regulator, in conjunction with an alternator, will satisfactorily maintain the battery in a state of charge over a wide range of operating conditions.

ALTERNATOR TESTS

ALTERNATOR TESTING
GENERAL MOTORS DELCOTRON

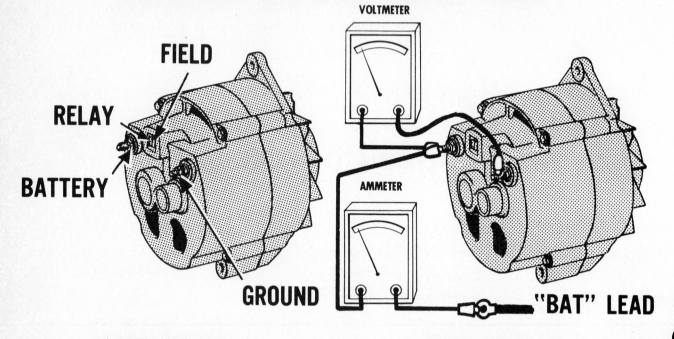

VOLTMETER

FIELD

RELAY

BATTERY

AMMETER

GROUND

"BAT" LEAD

TERMINAL
IDENTIFICATION

DELCOTRON
TEST HOOK-UP

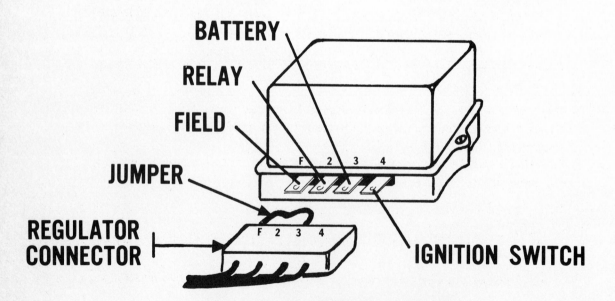

BATTERY

RELAY

FIELD

F 2 3 4

JUMPER

REGULATOR
CONNECTOR

F 2 3 4

IGNITION SWITCH

TWO-UNIT REGULATOR

ALTERNATOR TESTING

GENERAL MOTORS DELCOTRON
This procedure is for testing present models of Delcotron units using a 2-unit regulator which consists of a voltage regulator and a field relay. The charging system may employ either an indicator light or an ammeter.

DELCOTRON OUTPUT TEST

1972 models: remove air conditioner/heater motor fuse.

1. Check and adjust drive belt tension as required.

2. Disconnect battery ground cable.

3. Disconnect lead from Delcotron "BAT" terminal and connect ammeter between terminal and disconnected lead.

4. Reconnect battery ground cable.

5. Connect carbon pile rheostat across battery. Connect tachometer.

6. Connect voltmeter across Delcotron from "BAT" terminal to "GRD" terminal or across battery.

7. Turn on all electrical accessories, set parking brake firmly, start and idle engine with transmission selector in DRIVE position. Delcotron output should be at least 10 amperes.

8. Shift selector to PARK position and increase engine speed to 1500 rpm. Ampere output should be as specified. Adjust load rheostat, if required, to obtain desired output.
   Note: In the absence of a carbon pile rheostat, turning the headlights on high beam usually supplies the necessary load.

If output is less than specified, supply field current directly to Delcotron, by either of two methods, as follows:

Unplug Delcotron connector and connect jumper lead from "F" terminal to "BAT" terminal in connector or connect jumper lead from Delcotron "F" terminal to "BAT" (3) terminal. Repeat test in Step 8.
Caution: Increase engine speed slowly to prevent voltage from exceeding maximum specified limit.

   a. If output remains low, Delcotron is defective.

   b. If output rises to normal output, regulator or connecting wiring is at fault.

   c. Stop engine. Remove jumper lead and reconnect harness. Install fuse.

VOLTAGE REGULATOR TEST

1. Install ambient temperature gauge on regulator cover.

2. With all electrical loads OFF, run engine for 15 minutes at about 1500 rpm.

3. Note ammeter reading. For accurate voltage check, ammeter should read between 3 and 10 amperes.

   If ammeter reading is still high after 15 minute run, substitute a fully charged battery and proceed with test.

   Note: A ¼ ohm resistor in series with the ammeter will simulate a fully charged battery and will permit an uninterrupted test.

4. Momentarily increase engine speed to about 2000 rpm and note readings on voltmeter and temperature gauge. Check readings against specifications for voltage setting at regulator ambient temperature. Make note of amount of adjustment required to place setting in specified limits.

   a. Remove regulator cover. BE CAREFUL to lift cover straight up. If field relay or voltage coil are touched by cover, the resulting arc may ruin the regulator.

      Voltage reading will change somewhat when cover is removed.

   b. Increase or decrease voltage setting the amount previously noted. Always make final adjustment by slightly increasing the spring tension.

   c. Carefully replace cover and recheck voltage setting.

The spread in the voltage settings is provided to permit tailoring the setting to the specific requirements of the vehicle being tested. In any event, however, the final setting must be within the specified range.

ALTERNATOR TESTS

ALTERNATOR TESTING

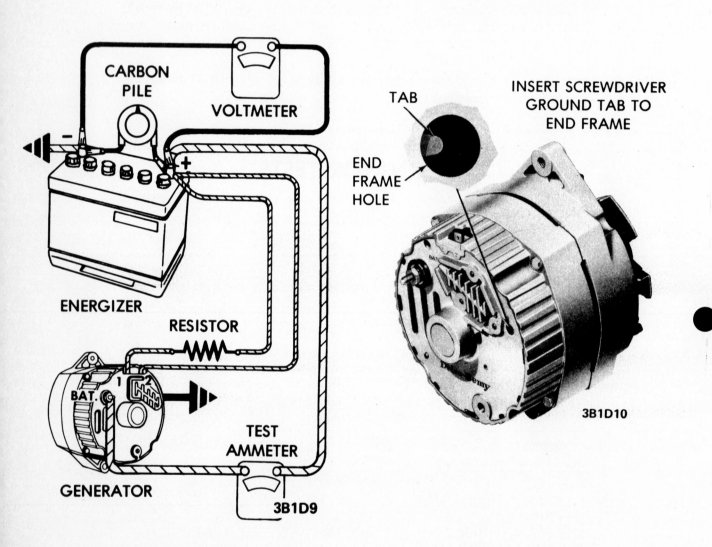

CARBON PILE

VOLTMETER

ENERGIZER

RESISTOR

GENERATOR

BAT.

TEST AMMETER

3B1D9

TAB

INSERT SCREWDRIVER GROUND TAB TO END FRAME

END FRAME HOLE

3B1D10

TESTING THE DELCOTRON WITH INTERNAL REGULATOR

ALTERNATOR TESTING

GENERAL MOTORS DELCOTRON—WITH INTEGRAL REGULATOR

This procedure is for testing Delcotron units that use an integral (internal) regulator and incorporate either an indicator light or an ammeter.

DELCOTRON OUTPUT TEST

Important: When testing, all accessories must be shut off and the blower motor lead disconnected.

1. Check and adjust drive belt tension as required.
2. As a safety precaution, disconnect battery ground cable while connecting test instruments.
3. Disconnect the lead from the "BAT" terminal and connect a test ammeter (0–75 Amp range or higher) between the disconnected wire and the "BAT" terminal.
4. Reconnect battery ground cable.
5. Operate engine at recommended test speed (typically 1500–2000 rpm).
6. Using a screwdriver, momentarily ground the tab to the Delcotron housing (see illustration) and observe the test ammeter. The reading should equal the specified output current ±10%. If so, the Delcotron is considered good. If not, the unit must be removed for bench testing and servicing.

   Note: Do not ground the tab any longer than is necessary to obtain an ammeter reading.

DELCOTRON VOLTAGE REGULATOR TEST

1. Connect a voltmeter (0–16 volt range) across the battery terminals. Leave ammeter connected as before.
2. With all electrical loads off, run engine for about 15 minutes at a fast idle. This is to normalize the system and, if necessary, partially recharge the battery.
3. At the specified test speed (usually 1500 rpm) note the ammeter reading. If it is less than 10 amperes, then check the voltmeter reading. This will be the voltage regulator setting, which should be within the specified limits (for example, 14.0 ±0.5 volts). The regulator is non-adjustable.
   If the ammeter shows more than 10 amperes, it indicates that the battery is too discharged for accurate voltage regulator testing. Install a fully charged battery and repeat the test.
   Note: Some charging system testers have provision for inserting a ¼ ohm resistor in the charging system to simulate a fully charged battery. This allows accurate tests of the voltage regulator with a discharged battery. Follow the instrument maker's prescribed test procedure.
4. An out-of-tolerance voltage regulator must be replaced. Since the regulator is internal, the Delcotron must be removed from the vehicle.

ALTERNATOR TESTING
CHRYSLER CORPORATION

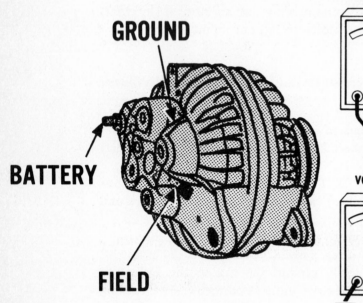

TERMINAL IDENTIFICATION

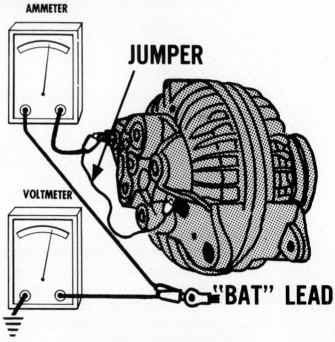

ALTERNATOR TEST HOOK-UP

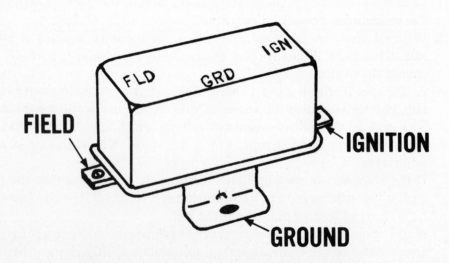

SINGLE-UNIT REGULATOR

ALTERNATOR TESTING

CHRYSLER CORPORATION

The regulator used on Chrysler Corporation cars contains only a voltage limiting relay.

Chrysler recommends that a fully charged battery be installed in the vehicle before charging system tests are conducted if the vehicle's battery is in a state of undercharge. Also make certain the alternator drive belt is properly tensioned.

ALTERNATOR OUTPUT TEST

1972 models: remove air conditioner/heater motor fuse.

1. Connect test ammeter between alternator "BAT" terminal and disconnected wire.

2. Connect jumper lead from alternator "Field" terminal to "BAT" terminal.

3. Connect voltmeter between disconnected "BAT" terminal lead and a good ground.

4. Connect tachometer.

5. Connect a carbon pile rheostat across the battery. Place rheostat in "Open" or OFF position.

6. Set engine speed to 1250 rpm. Adjust carbon pile to control alternator output voltage at 15.0 volts.

7. Observe ammeter for current output and compare reading to specifications. Output reading should be as specified with a plus or minus 3 amperes tolerance.

8. Stop engine. Remove jumper lead and rheostat. Install fuse.

VOLTAGE REGULATOR TEST

The regulator is checked by performing two tests. The first test checks the regulator's ability to maintain a specified voltage at low vehicle speed with heavy electrical loads. The second test checks the ability of the regulator to maintain voltage control at high vehicle speeds with light electrical loads.

FIRST TEST (Upper contacts)

1. Install ambient temperature gauge on regulator cover.

2. Set engine speed to 1250 rpm. Turn on lights and electrical accessories to obtain a 15 ampere charge rate. Operate engine for 15 minutes at this speed and load to normalize charging system temperature.

3. After 15 minutes reset engine speed to 1250 rpm. if required; readjust load to 15 amperes, as required.

4. Observe voltmeter and temperature readings and compare readings to specifications.

SECOND TEST (Lower contacts)
5. Increase engine speed to 2200 rpm.

6. Turn off all electrical accessories and observe ammeter and voltmeter.

Amperage should decrease and voltage should increase at least 0.2 volts but <u>not</u> more than 0.7 volts. Ampere output should be 5 amperes or less.

A regulator that does not conform to these specifications must be adjusted. It may be necessary to reset the armature air gap and contact point spacing. Readjust the voltage limiter by bending the spring hanger down to increase the voltage setting or up to decrease the setting. A regulator that cannot be brought into the specified operating range by adjustment, must be replaced.

ALTERNATOR TESTS

ALTERNATOR TESTING

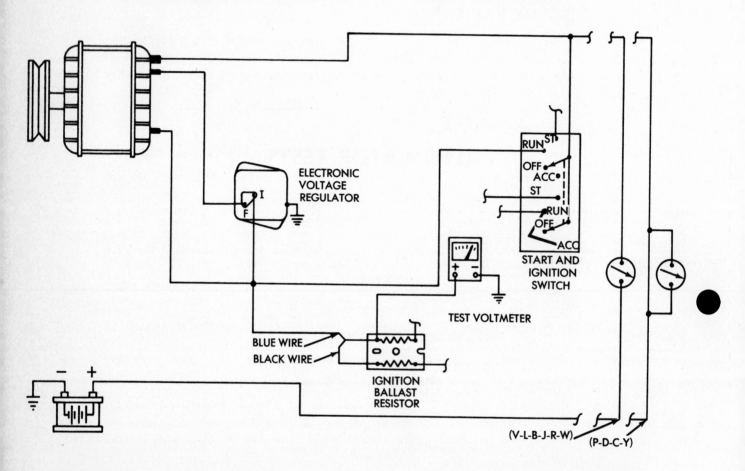

TESTING THE CHRYSLER ELECTRONIC VOLTAGE REGULATOR

ALTERNATOR TESTING

CHRYSLER CORPORATION

This procedure is for use with systems incorporating a solid state voltage regulator. Be sure the battery is near full charge before testing. Also check the drive belt tension and make certain the blower motor lead is disconnected. Turn all accessories off.

ALTERNATOR OUTPUT TEST

1. Disconnect the lead from the alternator "battery" terminal and connect a test ammeter (0–75 amp range or higher) in series with the disconnected wire and the "battery" terminal.

2. Connect a voltmeter (0–16 volt range) to the "battery" terminal and ground.

3. Connect a carbon pile across the battery terminals. Be sure the pile is **OFF**.

4. Disconnect the green field wire from its terminal at the alternator. Set it aside.

5. Connect a jumper lead from this field terminal on the alternator to ground.

6. Start the engine and let it *idle*.

7. Increase the speed while slowly turning in the carbon pile until 1250 rpm and 15.0 volts is obtained. The carbon pile will control the voltage, which should never be allowed to exceed 16.0 volts.

8. At the above speed and voltage, note the ammeter reading. This should be within the specified range. If not, remove the alternator for bench testing.

9. Reduce speed, turn off the carbon pile and remove jumper lead. Reconnect field lead.

Voltage Regulator Test (solid state unit)

1. Be sure battery is adequately charged.

2. Connect the voltage meter to the battery side of the ignition ballast resistor and ground. Be sure all accessories are off.

3. Run the engine at 1250 rpm and note voltmeter reading. Compare with specified voltage regulator range (typically, 13.8 and 14.4 at 80°F).

4. If the voltage reading is high, check for a poor regulator ground before replacing the regulator. If the reading is low, replace the regulator. The regulator is non-adjustable.

5. Stop engine and remove test instruments.

ALTERNATOR TESTING
FORD MOTOR COMPANY

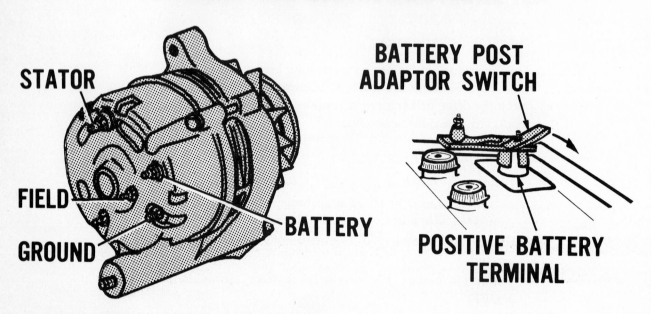

STATOR

FIELD

GROUND

BATTERY

BATTERY POST
ADAPTOR SWITCH

POSITIVE BATTERY
TERMINAL

TERMINAL IDENTIFICATION

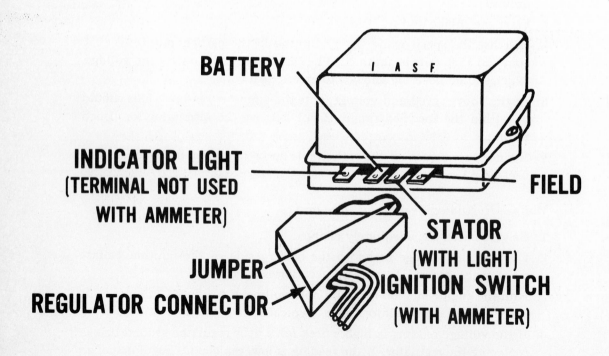

BATTERY

I A S F

INDICATOR LIGHT
(TERMINAL NOT USED
WITH AMMETER)

FIELD

STATOR
(WITH LIGHT)

JUMPER

IGNITION SWITCH
(WITH AMMETER)

REGULATOR CONNECTOR

TWO-UNIT REGULATOR

ALTERNATOR TESTING

FORD MOTOR COMPANY

Regulators used by Ford Motors contain two units, a voltage limiter and a field relay. The charging system may employ either an indicator light or an ammeter.

ALTERNATOR OUTPUT TEST

1972 models: remove air conditioner/heater motor fuse.

1. Remove ground and insulated battery cable clamps from battery.

2. Install Battery Post Adapter Switch on positive battery post. Place switch in "Open" position. Connect insulated battery cable to battery post adapter.

3. Disconnect plug connector at regulator and connect a jumper lead between the "A" (Battery) terminal and the "F" (Field) terminal of the connector.

4. Connect the tester ammeter leads to the Battery Post Adapter Switch, observing polarity. Turn tester control knob to the "Direct" position.

5. Reconnect battery ground cable to battery negative post.

6. Connect voltmeter leads across battery, observing polarity.

7. Connect tester ground lead to battery negative post.

8. Observe ammeter. A reverse current flow of from 2 to 3 amperes (field current) should be indicated.

9. Turn tester control knob to "Load" position.

10. Close Battery Post Adapter Switch and start engine. Open switch. Note: The Battery Post Adapter Switch is closed only during engine starting operations. It is open at all times tests are being conducted.

11. Slowly increase engine speed to about 2500 rpm while slowly rotating tester control knob to limit alternator output voltage to 15.0 volts.

12. Observe ammeter for alternator output and compare reading to specifications.

13. Stop engine. Remove jumper lead and reconnect connector. Install fuse.

VOLTAGE REGULATOR TEST

Leave Battery Post Adapter Switch and tester ammeter and voltmeter leads connected as in alternator output test. All lights and electrical accessories must be turned "Off".

1. Install ambient temperature gauge to regulator cover.
2. Close battery switch, start engine, open switch, set engine speed at 2000 rpm.
3. Rotate tester control knob to the ¼ ohm position. Ammeter should indicate about a 2 ampere current flow.
4. Observe voltmeter and regulator ambient temperature.

   Compare readings to specifications.

   A voltage reading out of the allowable range of the specifications should be adjusted to a setting midway in the range.

   A voltage reading within the limits of the specifications but with either an undercharged or overcharged battery condition, may be "tailored" to the individual vehicle's requirements, by raising or lowering the voltage setting slightly to compensate for the condition. In any event, the final setting must be within the range of the specification.

ON-THE-VEHICLE DIODE TEST

A low reading during the alternator output test may be caused by one or more defective diodes. This procedure provides a way of testing these alternator diodes without removing the alternator. Only a voltmeter is required.

1. Disconnect the electric choke lead, if used.
2. Disconnect the voltage regulator wiring plug and connect a jumper between the A and the F terminals of this plug (the same connections used for the output test).
3. Connect a voltmeter (0–16 volt range) across the battery terminals.
4. Operate the engine at idle speed.
5. Record the voltmeter reading.
6. Move the positive voltmeter lead to the S terminal of the regulator plug and note the voltage reading:
   a. If it is ½ of the reading in Step 5, the diodes are good.
   b. If it is more than 1.5 volts above or below ½ of the reading in Step 5, one or more defective diodes are indicated.
7. Be sure to reconnect the electric choke (if so equipped) after testing.

DIODE TESTS

DIODE TESTS

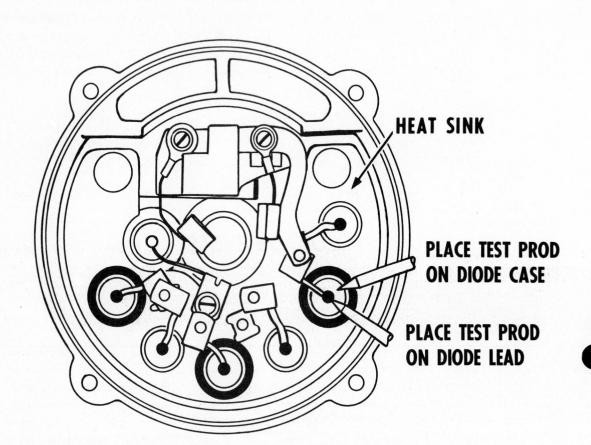

HEAT SINK

PLACE TEST PROD
ON DIODE CASE

PLACE TEST PROD
ON DIODE LEAD

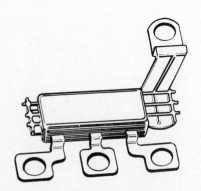

DIODE TRIO

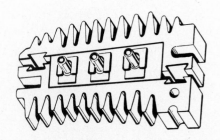

RECTIFIER BRIDGE

DIODE TESTS

Due to the manner in which the three stator windings are connected, two windings are always being used at any one time. Only two diodes are being used at any one time. If any diode is defective, causing one phase of the 3-phase winding to be missing, alternator output is decreased by approximately two-thirds because any single missing phase undesirably influences both of the other phases.

The diodes on some makes of alternators can be tested with the alternator assembled and on the engine. Other makes have to be disassembled for diode testing.

Diodes can be readily checked using a Diode Tester. When using this test instrument follow the manufacturer's instructions.

Diodes may also be tested with a **12-volt** test lamp. Touch the prods of the test lamp leads to the diode case and diode lead and then reverse the test lamp prods as previously explained. If the diode is good, the test lamp will light in only one test. If the lamp fails to light in both tests or lights in both tests, the diode is defective.

Another method of checking diodes is with an ohmmeter. After the stator leads have been disconnected, each diode can be tested for shorts and opens. Touch one ohmmeter prod to the diode case and the other prod to the diode lead and observe the ohmmeter reading. Then reverse the ohmmeter prods and again observe the meter readings. A diode in good condition will have one high reading and one low reading. If both readings are very low, or if both readings are very high, the diode is defective. Push and pull the diode lead **gently** while testing it to detect loose connections. Test all six diodes in the same manner. The readings of the diodes in the insulated heat sink will be opposite from those in the grounded heat sink or end frame.

When a diode is found defective, it is advisable to replace all three diodes in the end frame or heat sink using the correct procedure and the proper removal and installation tools. Diodes must never be hammered into position as the impact can easily crack the silicon wafer. Be sure to test the condenser as it may also be damaged.

Recently, the alternator rectifier bridge was introduced. The bridge contains all 6 diodes. This new design further simplifies the diode replacement operation. The bridge also contains the fins that are necessary for cooling the heat sink in which the diodes are mounted.

THE DIODE TRIO

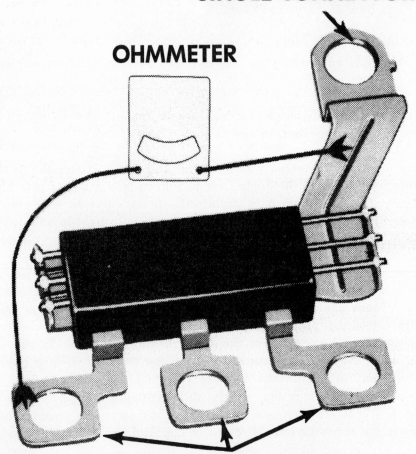

SINGLE CONNECTOR

OHMMETER

THREE CONNECTORS

TESTING A DIODE TRIO

THE DIODE TRIO

Some late model alternators use, in addition to the regular diode assembly or rectifier bridge, a diode trio. It consists of three diodes contained in a single package, as shown, and is used to supply field current to the rotor. This unit is mounted internally in the alternator and must be removed for testing.

Testing is best accomplished with an ohmmeter as shown in the illustration. Connect one ohmmeter lead to the single connection on the end and then touch each of the three other connections. The reading should all be the same, either infinitely high or very low. Then reverse the ohmmeter leads and repeat the test. These readings should be the opposite of the first group. If the ohmmeter gives the same reading with both connections, the diode trio is defective.

FIELD WINDING TESTS

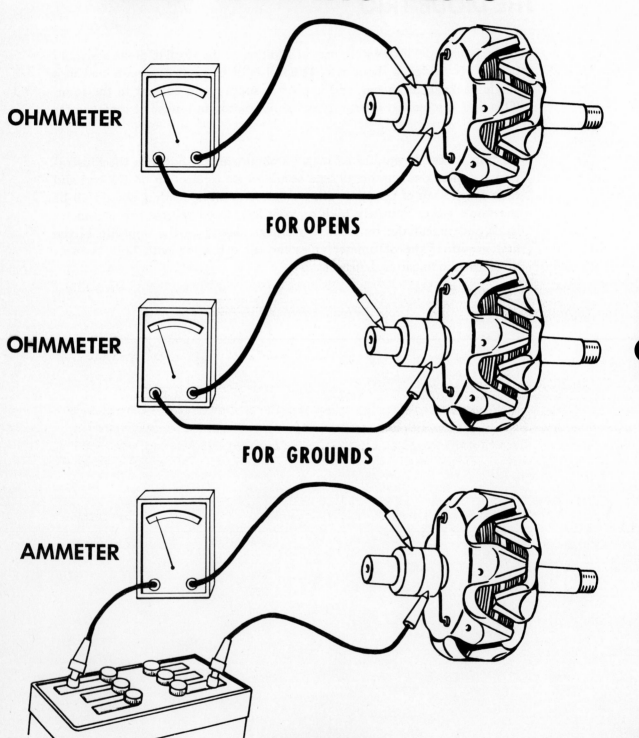

OHMMETER

FOR OPENS

OHMMETER

FOR GROUNDS

AMMETER

FOR SHORTS

FIELD WINDING TESTS

The rotor field winding may be checked electrically for open circuit, ground circuit and short circuit.

To check for open circuit, touch a 110-volt test lamp prod to each slip ring. If the lamp fails to light, the field winding is open.

To check for ground circuit, touch one 110-volt test lamp prod to either slip ring and the other test lamp prod to the rotor shaft. If the lamp lights, the field winding is grounded.

To check for short circuit, connect a 12-volt battery and an ammeter in series with the two slip rings. The ammeter should indicate approximately 2 amperes. An ammeter reading above this value indicates a shorted field coil winding. High output alternators will have a higher field current draw.

If an ohmmeter is available, these three tests can be conducted by measuring resistance values.

STATOR WINDING TESTS

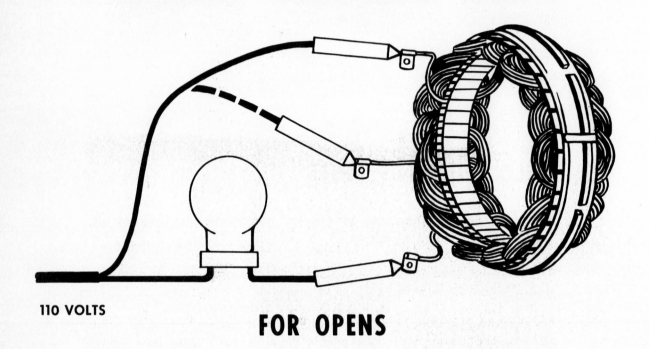

110 VOLTS

FOR OPENS

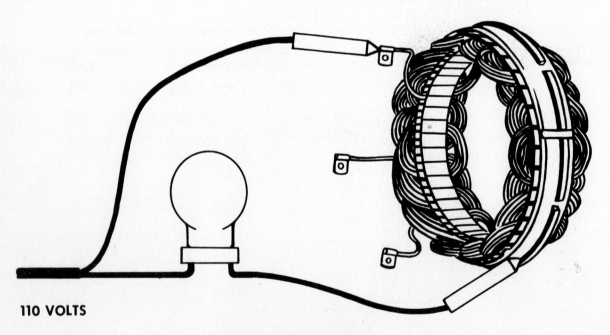

110 VOLTS

FOR GROUNDS

STATOR WINDING TESTS

Stator windings are checked for open circuit, ground circuit, and short circuit.

To check for open circuit, touch the prods of a 110-volt test lamp to the stator winding terminals as illustrated. If the lamp does not light, the stator winding is open. To complete the test, move one test lamp prod to the other stator winding terminal.

To check for a ground circuit, touch one prod of the 110-volt test lamp to the stator frame and the other test prod to any of the stator winding terminals. The test lamp should not light. If it does, the stator winding is grounded.

If an ohmmeter is available, these tests may be conducted by measuring resistance values.

A visual inspection for charred winding insulation should also be conducted at this time.

AC SYSTEM
SERVICE PRECAUTIONS

1 - Always be *absolutely sure* that the battery ground polarity and the charging system polarity are the same, when installing a battery.

2 - *Do not* polarize an alternator.

3 - *Never* short across or ground any of the terminals on either the alternator or the regulator.

4 - *Do not* operate an alternator on open circuit.

5 - Booster battery *must be* correctly connected.

6 - Battery charger *must be* correctly connected.

7 - *Always* disconnect the battery ground cable before replacing or servicing electrical units.

AC SYSTEM SERVICE PRECAUTIONS

Alternators are designed and constructed to give long periods of trouble-free service with minimum maintenance. To avoid accidental damage to the alternator, regulator or charging system wiring, the following precautions should be observed:

1. Always be **absolutely sure** that the battery ground polarity and the charging system polarity are the same, when installing a battery.

   If a battery is hooked-up backwards, it is directly shorted across the alternator diodes. The high current flow can damage the diodes and even burn up the wiring harness. If battery post identification is not obvious, use a voltmeter across the posts to identify their polarity.

2. **Do not** polarize an alternator.

   The reason a DC generator is polarized is to excite the generator field to insure that the generator and battery will have the same polarity. Since the alternator develops voltage of both polarities, which the diodes automatically rectify, there is no need to polarize an alternator. In fact, damage to the alternator, regulator or circuits may result from an attempt to polarize the alternator.

3. **Never** short across or ground any of the terminals on either the alternator or the regulator.

   Care should be exercised when working in the engine compartment to avoid accidental shorting of the alternator or regulator terminals. Shorting or grounding of the alternator or regulator terminals, either accidental or deliberate, can result in damage to the diodes, the regulator and/or the wiring. Grounding of the alternator output terminal (Bat) even when the engine is not running can result in damage since battery voltage is applied to this terminal at all times. Care should also be exercised when adjusting the voltage regulator to prevent accidental shorting.

4. **Do not** operate an alternator on open circuit.

   Operating the alternator while it is not connected to the battery or to any electrical load will cause the voltage developed to be extremely high. This high voltage can damage the diodes.

5. Booster battery **must be** correctly connected.

When the booster battery is used to assist in engine starting, it must be connected to the car battery in proper polarity to prevent damage to the diodes. The positive cable from the booster battery must be connected to the car battery positive terminal and the negative cable from the booster battery must be connected to the car battery negative terminal. Positive to positive and negative to negative is the proper hookup.

6. Battery charger **must be** correctly connected.

Battery charger leads must be correctly connected to the battery, the positive charger cable to the positive battery post and the negative charger cable to the negative battery post. Failure to observe this precaution may also result in damage to the diode rectifiers.

When charging a battery, disconnect the battery cables before connecting the charger leads to the battery to prevent possible damage to the alternator.

A fast battery charger should **never** be used as a booster for starting the engine in a car equipped with an alternator.

7. **Always** disconnect the battery ground cable before replacing or servicing electrical units.

Disconnecting the battery ground cable is always advisable when replacing electrical units or servicing electrical components. This precaution will prevent accidental shorting which may result in damage to the diodes, regulator or wiring. Remember, too, that if the battery ground cable is not disconnected, **make sure** the ignition switch is turned off when servicing the regulator since the alternator field circuit is connected to the battery through the ignition switch.

THE
IGNITION SYSTEM

THE IGNITION CIRCUIT

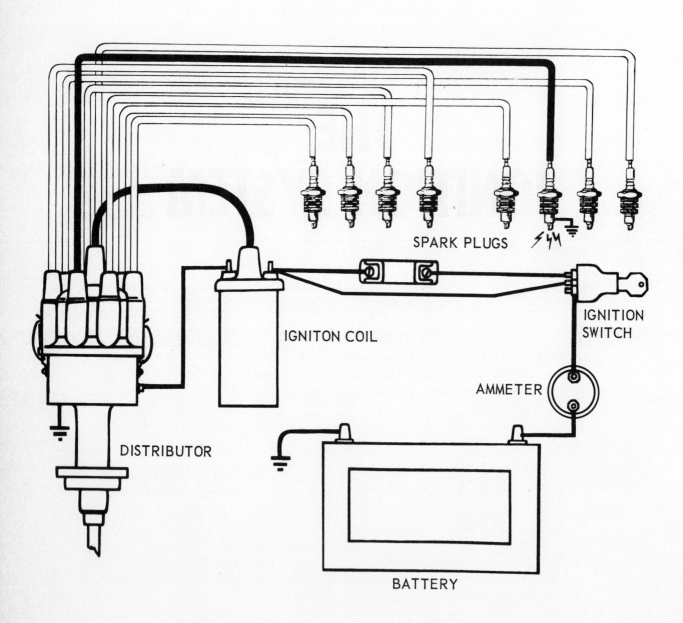

SPARK PLUGS

IGNITON COIL

IGNITION
SWITCH

AMMETER

DISTRIBUTOR

BATTERY

THE IGNITION CIRCUIT

The internal combustion engine operates through the forces created by expanding gases in the combustion chambers. These gases are the product of the burning air-fuel mixture which was ignited by a high voltage spark.

The function of the ignition system is to produce a high voltage surge, and to deliver it to the proper spark plug at the correct instant to ignite the air-fuel mixture compressed in the cylinder.

The ignition system consists of the following components:

Battery	Distributor Assembly
Ignition Switch	Body Breaker Points
Ballast Resistor	Cap Condenser
or Ballast Wire	Rotor Advance Mechanisms
Ignition Coil	High Tension Leads
	Spark Plugs

The battery (or generator) is the source of power which supplies low voltage to produce a current flow in the ignition primary circuit.

The ignition switch is simply an On-Off switch to complete the ignition circuit between the battery and coil. When the ignition switch is closed, current will flow through the coil-distributor (primary) circuit and return by way of the car frame or engine block to the battery. The ignition switch also serves as a bypass switch during engine starting.

The ballast resistor in the ignition primary circuit is designed to permit the proper amount of current flow for all driving conditions. During cranking, however, it is bypassed to permit full battery voltage and maximum current flow through the coil for quick starting.

The ignition coil transforms or "steps up" the low battery voltage to a voltage high enough to jump a spark gap at the spark plug.

The distributor interrupts the flow of current in the primary winding by the action of the breaker points. It also distributes the high-tension current developed by the coil through the rotor and the distributor cap to the proper spark plug at the correct instant.

The high tension or secondary leads conduct the high voltage produced by the ignition coil to the distributor, and from the distributor to the spark plugs.

The spark plugs provide a spark gap in the combustion chamber. When high voltage jumps across this gap, the air-fuel mixture is ignited.

THE IGNITION SYSTEM
(IGNITION RESISTOR BY-PASS
IGNITION AND STARTING SWITCH TYPE)

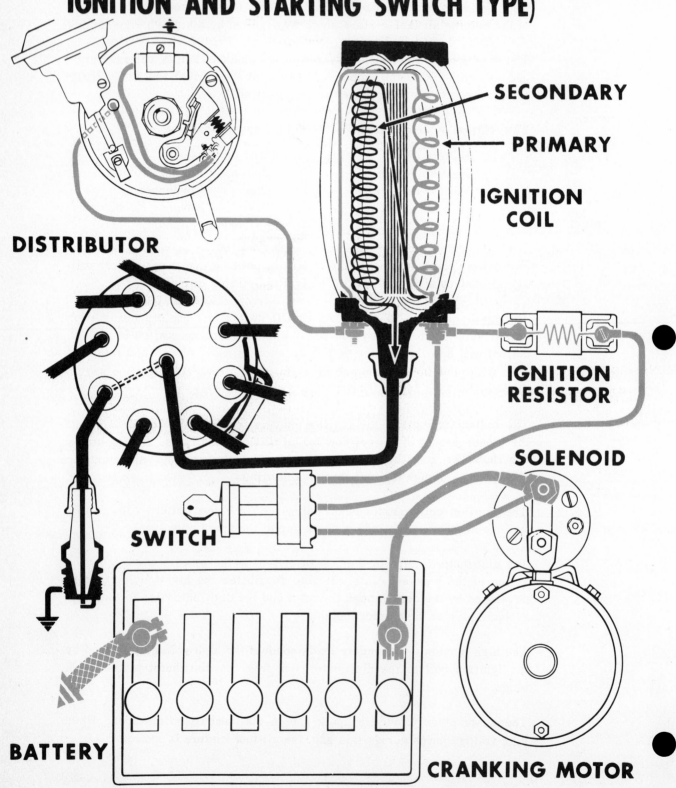

SECONDARY

PRIMARY

IGNITION COIL

DISTRIBUTOR

IGNITION RESISTOR

SOLENOID

SWITCH

BATTERY

CRANKING MOTOR

IGNITION SYSTEM OPERATION

The ignition primary circuit is also called the low voltage or the low tension circuit. This is the circuit through which current flows at battery or generator voltage. The ignition switch, resistor, ignition coil primary winding, breaker points and the condenser are all in the primary circuit.

The secondary ignition circuit is also called the high voltage or high tension circuit. The coil secondary winding, distributor rotor, distributor cap, high tension leads, and the spark plugs are all part of the secondary circuit.

When the ignition switch is turned on and the breaker points are closed, current flows in the primary circuit. As current flows through the primary winding of the ignition coil, a strong magnetic field is produced in the coil. When the breaker points open, current through the primary winding of the coil is stopped, and the magnetic field around the coil winding collapses. These collapsing lines of force cut across both the primary and secondary windings inducing a very high voltage in the secondary winding. The high voltage so induced forces current to jump the spark plug gap.

The primary ballast resistor is essentially a current compensating device consisting of a resistor unit or wire located in the primary ignition circuit. The compensating action is obtained because, at low engine speeds, the current flows for longer periods of time. This heats up the resistor, thereby raising its resistance and reducing current flow. This action serves to keep the coil primary winding cooler and improves distributor breaker point life. At high speeds, the current flows for shorter periods of time which lets the resistor cool and increases the current flow in the primary winding of the coil. This action permits maximum secondary voltage to be obtained.

Because of the lowered battery voltage resulting from the starter load on some vehicles, the ballast resistor is bypassed while the starting system is in operation. This is done to provide higher secondary voltage for starting.

DISTRIBUTOR ASSEMBLY

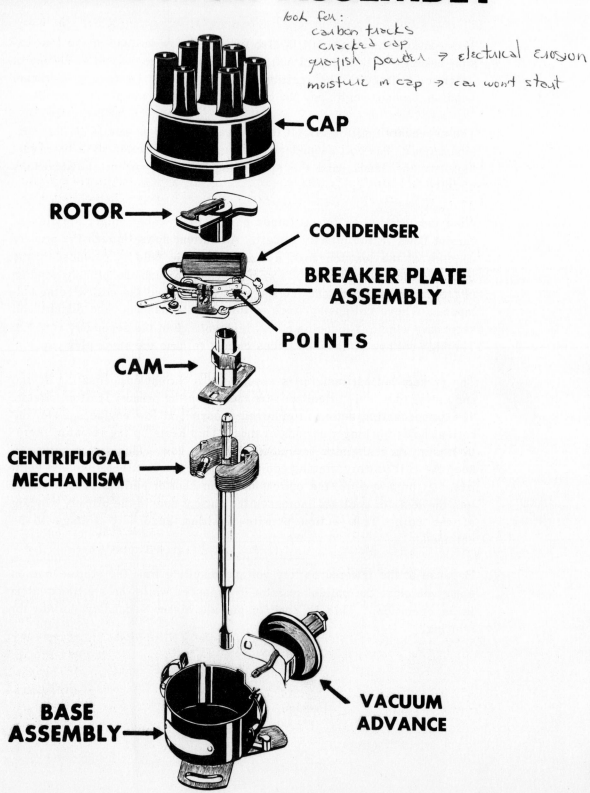

look for:
carbon tracks
cracked cap
grayish powder → electrical erosion
moisture in cap → car won't start

CAP

ROTOR

CONDENSER

BREAKER PLATE ASSEMBLY

POINTS

CAM

CENTRIFUGAL MECHANISM

VACUUM ADVANCE

BASE ASSEMBLY

DISTRIBUTOR ASSEMBLY

The distributor assembly is made up of several subassemblies: the cap, rotor, breaker plate assembly supporting the breaker points and condenser, cam, centrifugal mechanism, vacuum advance unit and the distributor base or body.

The distributor cap and rotor are used to distribute the high voltage current developed in the coil to each spark plug in firing order sequence. The distributor cap is keyed to the distributor body to maintain the correct relationship between the distributor cap towers and the rotor. The rotor is keyed to the distributor cam to be in the correct position to transfer the high voltage current to one of the distributor cap high voltage towers when the points open.

The breaker plate supports the breaker points at the correct position for the cam to open them. In most distributors, the breaker plate is moveable and is mounted on a center bearing or side pivot. Since a moving part can develop high resistance at the bearing surface, a flexible insulated wire is used to connect the points to the primary circuit. Also, a grounding wire is used between the plate and distributor housing.

The distributor cam opens the breaker points to interrupt the primary circuit. The cam has the same number of lobes as the engine has cylinders as the points must be opened to produce a high voltage surge for each power stroke. The distributor cam is actuated by the weights of the centrifugal advance mechanism.

The centrifugal advance mechanism varies the position of the cam in relation to engine speed. The mechanism has weights that are thrown out against calibrated springs to advance the breaker cam as speed increases. The advance mechanism is usually mounted in the distributor housing below the breaker plate. Some late model distributors have the centrifugal advance mechanism above the breaker plate and below the rotor.

The vacuum advance unit advances the ignition timing in addition to the advance provided by the centrifugal unit. The vacuum unit is controlled by manifold vacuum and consists of a metal chamber in which a flexible air-tight diaphragm is located. A link extends from one side of the diaphragm to the breaker plate assembly or distributor housing. If the distributor is equipped with a movable breaker plate, the vacuum advance mechanism rotates the plate to advance the ignition timing. If the plate is rigidly mounted in the distributor, the vacuum advance unit rotates the complete distributor assembly for advance.

1) #1 cyl. on comp. stroke
2) timing mark at TDC
3) rotor points to #1 terminal or plug wire

The lower drive end of the distributor shaft is equipped with either an offset grove or tang or with a gear, depending on the manner in which the distributor is driven. The reason the tang is offset is to assist in correct installation of the distributor when replacing it in the engine.

Regardless of the method of drive all distributors are driven at camshaft speed which is one-half engine speed.

The base assembly or body serves as a housing for all the parts mentioned and also serves as the bearing for the distributor shaft. The bearing may be cast iron, bronze bushing or ball bearings. The base is also the adapter for installing the distributor assembly in the engine block. The base can be rotated in either direction to make the initial timing adjustment.

DWELL ANGLE

DWELL ANGLE

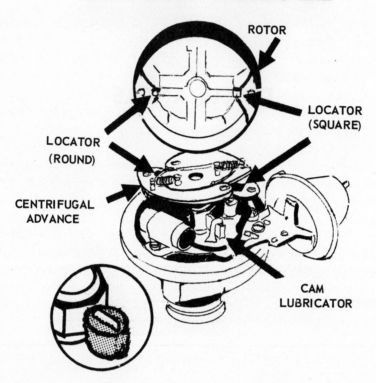

ROTOR

LOCATOR (SQUARE)

LOCATOR (ROUND)

CENTRIFUGAL ADVANCE

CAM LUBRICATOR

CONTACT POINT DWELL TEST

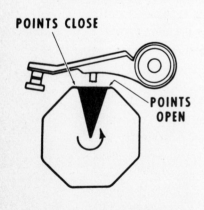

POINTS CLOSE

POINTS OPEN

NORMAL DWELL-NORMAL GAP

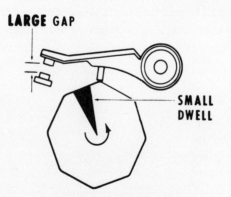

LARGE GAP

SMALL DWELL

INSUFFICIENT DWELL

SMALL GAP

LARGE DWELL

EXCESSIVE DWELL

DWELL ANGLE

Dwell angle is the number of degrees the distributor cam turns during which time the breaker points are closed. During the dwell period, a magnetic field is built up in the primary winding of the coil. However, time is required to build up a full strength magnetic field. When a full strength magnetic field is produced, the coil is said to be "saturated." To be assured of coil saturation at low engine speeds presents no problem because of the relatively slow rotation of the distributor cam. At high speeds, however, unless the distributor points are adjusted to provide a sufficient dwell period, coil saturation will not be attained.

In a six-cylinder engine running at idle speed of 400 rpm, the ignition system must produce 20 sparks per second to fire all of the cylinders. With an engine operating at this speed, dwell angle is not very critical because more than sufficient coil "saturation" time is available. However, with the engine running at 4000 rpm, or the equivalent of 90 miles per hour or more, 200 sparks per second would be required to fire all the cylinders. This is the speed at which dwell becomes extremely critical. If the dwell was reduced only slightly from the required amount, the engine would begin to misfire at high speed because the coil does not have time to become sufficiently "saturated".

Although dwell is not critical at low speed, point gap becomes very important. With the engine cranking, there must be sufficient point gap or the points will arc excessively and the engine will not start readily. Also, if an engine is operated with too little point gap at low speed, the points will deteriorate rapidly. If the points open slowly, and do not open wide enough, an arc will continue across the contact points using energy that would normally create a spark at one of the spark plugs. When an arc occurs the engine usually misfires because the energy of the primary circuit is dissipated preventing sufficient secondary circuit voltage build-up.

Dwell angle adjustment directly affects ignition timing. Under certain conditions the rubbing block on the movable breaker arm may wear. As a result, the dwell angle increases which in turn causes the ignition timing to be late.

One of the largest single causes of breaker point failure is the lack of cam lubricant. Point rubbing block wear can be appreciably reduced by applying a thin film of high-temperature cam lubricant to the distributor cam when servicing the distributor. It is important that the proper lubricant be used since it must be able to: adhere to the cam surface at high cam speed; resist melting at high temperature; resist chemical reaction with the polished steel cam; effectively control moisture to prevent rust formation on the cam; resist drying out with age.

Some distributors are fitted with a cam lubricator the wick of which is impregnated with a special lubricant. At specified intervals the cam lubricator should be rotated 180° (or end-for-end) with the lubricator just touching the cam lobes. At every other service interval the lubricator should be replaced. Lubricant should never be added to the lubricator or used on the cam of a distributor equipped with this device.

Ignition point spring tension plays an important part in the performance of the ignition system, and must be within specified limits. Excessive pressure causes rapid rubbing block and cam wear, while insufficient pressure will permit high-speed point bounce which, in turn, will cause arcing and burning of the points and misfiring of the engine.

SIDE-PIVOTED BREAKER PLATE

Before testing and adjusting the dwell angle on distributors that have a side-pivoted breaker plate, as on Chrysler, Ford, Autolite and Prestolite distributors, it is recommended the distributor vacuum line(s) be disconnected and plugged.

On some 1970-1972 Chrysler-built 8 cylinder engines equipped with a distributor solenoid, the solenoid lead should also be disconnected at the connector when testing the dwell angle. The reason for this specific procedure is that the dwell angle specification is based on "No vacuum advance" and "No solenoid retard." Therefore the vacuum line and the distributor solenoid must be disconnected. DO NOT attempt to disconnect the lead at the solenoid.

The above procedures are required because the distributor shaft and the cam rotate about the center point of the shaft. The breaker points, however, are mounted on the breaker plate which pivots about its anchor. The points and the cam are therefore moving about two different centers or pivots. If the dwell angle is tested or adjusted while the vacuum lines are connected, the breaker plate is rotated for spark advance causing the breaker gap to increase. This action results in a decreased dwell angle because the points are closed for a shorter time.

This is also the reason why the permissible dwell variation on this type distributor is approximately double that allowed for the center-pivoted plate type.

To accurately test and set the dwell angle on this type distributor it is important that the recommended procedure be followed.

Remember this — the selection of a quality precision-manufactured set of breaker points, their proper installation and precise adjustment is the most critical single factor in the efficient operation of the ignition system.

DUAL BREAKER POINTS

DUAL BREAKER POINTS

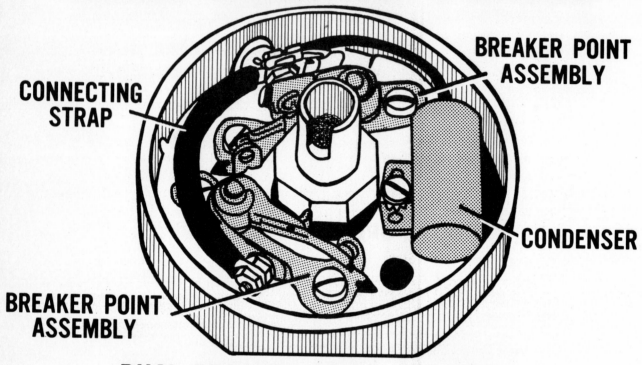

CONNECTING STRAP

BREAKER POINT ASSEMBLY

CONDENSER

BREAKER POINT ASSEMBLY

DUAL BREAKER POINT ASSEMBLY

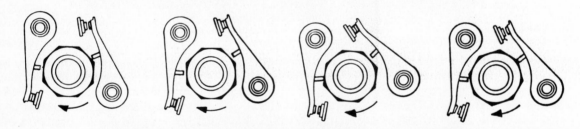

DUAL BREAKER POINT ACTION

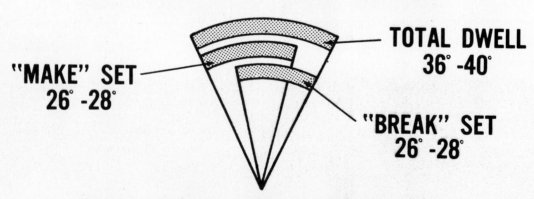

"MAKE" SET 26° -28°

TOTAL DWELL 36° -40°

"BREAK" SET 26° -28°

DUAL POINT DWELL OVERLAP

DUAL BREAKER POINTS

The function of dual breaker points is to provide a longer dwell period at high speed than is possible with a single breaker point set. The longer dwell period permits greater coil magnetic saturation with a proportionately higher secondary voltage output. This arrangement helps prevent the ignition system from "running out of spark", so to speak, during high speed operation. Another important factor relative to the use of dual points is that the increased dwell is accomplished while proper point gap is maintained.

One set of the dual points is called the "make" set and its function is to close the primary circuit and begin the dwell period. The other point set is called the "break" set and its function is to open the primary circuit which induces the high voltage generated in the coil secondary windings. Each point set is adjusted for a normal dwell period, say 26° to 28°. But since the dwell periods overlap each other slightly, the total dwell of both point sets may be 36° to 40°. It would be impossible to adjust a single point set to 40° dwell angle because the point gap would necessarily be so small that sparking would constantly occur across the points. Arcing across the points promotes pitting which interferes with quick starting and promotes premature point failure.

In the dual point set, an insulated jumper strap connects the two point sets in parallel. A single condenser is used. With the point sets connected in parallel, primary circuit current can flow through one point set, or the other, or both sets at the same time.

The action of dual points can be understood by studying the chart. The point set on the left of the cam is the "make" set. The point set on the right of the cam is the "break" set. Cam rotation is clockwise.

- The first illustration shows the "make" set just closing while the "break" set is still open. The primary circuit has just been completed and the dwell period begins.

- In the second illustration the "break" set has also closed. Both point sets are now closed and primary current flows through both sets further increasing the ignition coil magnetic field.

- The third illustration shows the "make" point set just starting to open. The primary circuit is not interrupted, however, because the "break" points are still closed.

- In the fourth illustration the "break" point set has opened. Both point sets are now open and the primary circuit is interrupted. The opening of the "break" set is quickly followed by the closing of the "make" set and another extended dwell period begins.

Dual breaker point adjustment will be covered in detail in the Tune-Up Procedure Section of this course.

AS ENGINE SPEED INCREASES SPARK MUST BE TIMED EARLIER

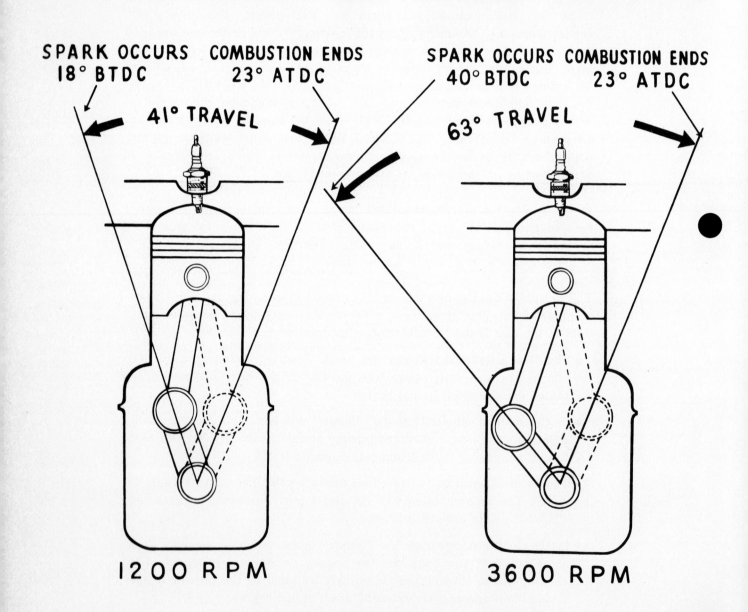

SPARK OCCURS 18° BTDC COMBUSTION ENDS 23° ATDC

41° TRAVEL

SPARK OCCURS 40° BTDC COMBUSTION ENDS 23° ATDC

63° TRAVEL

1200 RPM

3600 RPM

SPARK ADVANCE TIMING

To obtain full power from combustion, the maximum pressure must be reached just as the piston passes Top Dead Center, and combustion must be completed by approximately 23 degrees After Top Dead Center.

The fuel mixture ignited in the combustion chamber does not explode. It burns rapidly until the fuel is consumed. The time required for complete combustion is a small fraction of a second. For this reason ignition must take place before the piston passes Top Dead Center.

As engine speed increases, the piston moves through the compression stroke more rapidly, but the burning rate of the fuel mixture remains virtually the same. To compensate for the higher piston speed, ignition must occur earlier in the compression stroke.

During the combustion process at 1200 rpm, the crankshaft travels through 41 degrees of rotation; from the point of ignition, to 23 degrees After Top Dead Center. The spark must occur at 18 degrees Before Top Dead Center. The same engine running at 3600 rpm will require 63 degrees of crankshaft rotation to complete combustion by 23 degrees After Top Dead Center. This would require that the spark occur 40 degrees Before Top Dead Center. This is the reason why spark advance is such an important factor in efficient engine operation.

Setting the basic initial timing will be covered in detail in the Tune-Up Procedure Section of this course.

DISTRIBUTOR ASSEMBLY CENTRIFUGAL TYPE MECHANISM

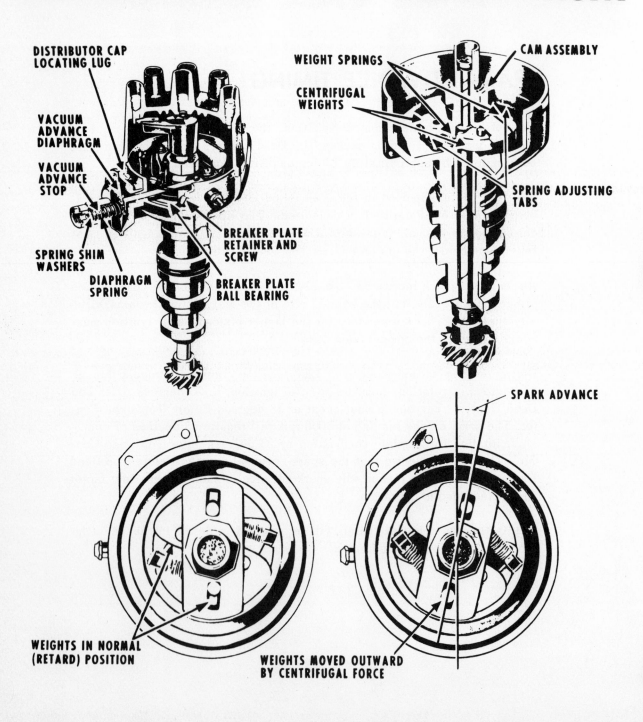

DISTRIBUTOR CAP LOCATING LUG

VACUUM ADVANCE DIAPHRAGM

VACUUM ADVANCE STOP

SPRING SHIM WASHERS

DIAPHRAGM SPRING

BREAKER PLATE RETAINER AND SCREW

BREAKER PLATE BALL BEARING

WEIGHT SPRINGS

CAM ASSEMBLY

CENTRIFUGAL WEIGHTS

SPRING ADJUSTING TABS

SPARK ADVANCE

WEIGHTS IN NORMAL (RETARD) POSITION

WEIGHTS MOVED OUTWARD BY CENTRIFUGAL FORCE

CENTRIFUGAL ADVANCE MECHANISM

Most distributors are equipped with a centrifugal-type advance mechanism. This mechanism automatically advances spark timing in correct relation to any engine speed above idle. The distributor cam is linked to the distributor shaft through the advance mechanism in such a manner that as distributor speed increases, the weights move out due to centrifugal force. This causes the cam to rotate several degrees ahead of the shaft, opening the breaker points earlier in the compression stroke thereby advancing the ignition timing. Springs return the weights to their retarded position as engine speed is decreased. This type of advance mechanism is actuated only by engine speed.

The amount of advance required at certain speeds will vary depending upon engine design, compression ratio, air-fuel ratio, and the octane rating of the fuel. Individual advance curves for different engines are designed into the contour of the cam and its slots and by using weight return springs of various lengths and tensions.

A quick method of checking if the centrifugal advance mechanism is working is to twist the rotor in the direction of rotation and then quickly release it. The rotor should snap back into its released position. Failure to return to its released position indicates broken return springs. A slow sluggish return action indicates a gummed or rusted condition of the advance mechanism. Either condition must be corrected before a tune-up can be effective.

Another method of checking the action of the centrifugal mechanism is to disconnect the vacuum line and tape its opening and slowly accelerate the engine while observing the timing mark position with a power timing light. If the mechanism is functioning, the timing mark will move against engine rotation as engine speed is increased.

VACUUM ADVANCE MECHANISM

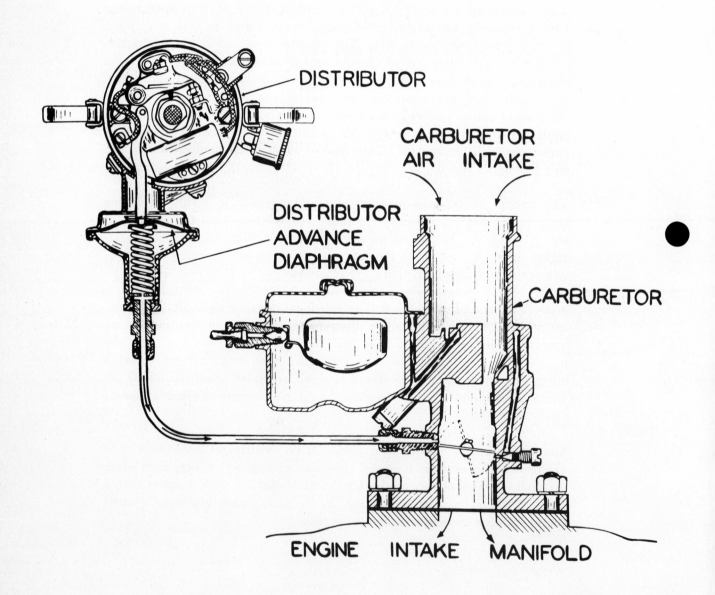

DISTRIBUTOR

CARBURETOR
AIR   INTAKE

DISTRIBUTOR
ADVANCE
DIAPHRAGM

CARBURETOR

ENGINE   INTAKE   MANIFOLD

VACUUM ADVANCE MECHANISM

During normal operation an engine does not always operate at full load. When operating at part load, the engine can operate at greater spark advance than when under full load. The additional advance during part throttle operation provides better economy, but usually does not have any noticeable effect on power and performance. This additional advance for part throttle operation is obtained by the use of a vacuum control mechanism operated by intake manifold vacuum.

When the engine is operating at light load, the throttle is partially closed and intake manifold vacuum will be high. When the engine is operating under heavy load, the throttle will be open and the manifold vacuum will be low. In other words, the greater load on the engine, the wider the throttle must be opened and the lower the manifold vacuum will become. Manifold vacuum is thus relative to engine loading.

The vacuum advance mechanism consists of a metal chamber in which a flexible, airtight diaphragm is located. A link extends from one side of the diaphragm to the breaker plate assembly or the distributor housing. A spring exerting pressure on the other side of the diaphragm causes the diaphragm to be flexed in such a way that the link holds the breaker plate in its retarded position. When vacuum is applied to the spring side of the diaphragm, it overcomes the spring pressure and moves the link which rotates the breaker plate against the distributor rotation resulting in an advance in ignition timing.

PORTED SPARK CONTROL

To allow manifold vacuum to be applied to the spring side of the diaphragm, the vacuum line leads from the diaphragm chamber to the base of the carburetor. The carburetor throttle plate acts as a dividing line or partition between atmospheric pressure and manifold vacuum. As the vacuum unit should be inoperative while the engine operates at idle speed, the vacuum line is coupled to the carburetor at a point just above the edge of the throttle plate when it is in a closed position. Under this condition, the weak carburetor vacuum cannot overcome the vacuum unit spring tension and the breaker plate remains in the full retard position. This arrangement is known as Ported Spark Control.

As the throttle is opened, the throttle valve will uncover the vacuum port and manifold vacuum will be impressed on the vacuum advance unit. The strong vacuum force will draw the diaphragm against its spring tension and rotate the breaker plate against distributor cam rotation. This action places the ignition point bumper block in a position where the cam lobe will contact it earlier in the compression stroke. By this action the ignition timing is advanced by the vacuum unit.

On some engines, depending on carburetor design, the vacuum opening to the carburetor may be below the throttle plate. In this case, the distributor will have some vacuum spark advance at idle speeds.

The amount of advance provided will depend upon the balance between intake manifold vacuum and the tension of the diaphragm spring. As the throttle is opened wider, the intake manifold vacuum will decrease and the diaphragm spring will cause the breaker plate to move toward the retarded position. At wide open throttle, manifold vacuum is practically zero and the vacuum advance mechanism will be in full retard position. The number of degrees of advance obtained from the vacuum advance mechanism for any given amount of manifold vacuum, will depend upon the tension of the diaphragm spring.

In some distributors the diaphragm spring tension may be adjusted by adding or removing shims which backup the diaphragm spring. Nonadjustable vacuum advance units must be replaced when tests indicate incorrect diaphragm spring tension.

Vacuum advance may be obtained in two ways; by rotating the vacuum breaker plate inside the distributor, or by rotating the entire distributor housing. The type of advance mechanism which rotates only the breaker plate is usually referred to as an internal vacuum advance mechanism. The type which rotates the entire distributor housing is usually referred to as an external vacuum advance mechanism.

TOTAL SPARK ADVANCE

SPARK ADVANCE

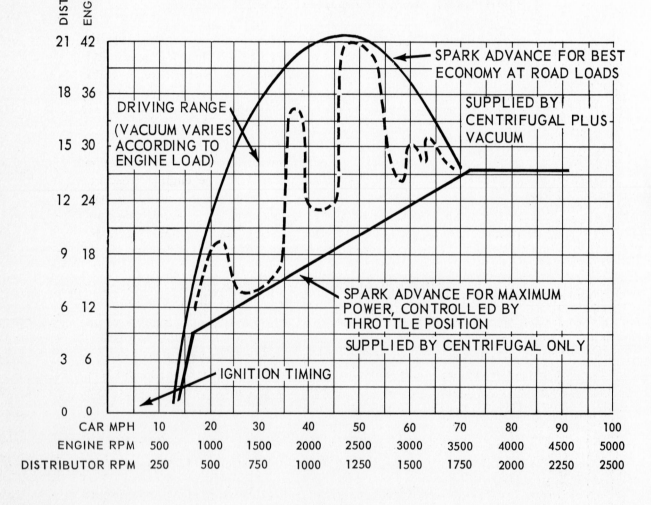

DISTRIBUTOR DEGREES

ENGINE DEGREES

SPARK ADVANCE FOR BEST ECONOMY AT ROAD LOADS

DRIVING RANGE (VACUUM VARIES ACCORDING TO ENGINE LOAD)

SUPPLIED BY CENTRIFUGAL PLUS VACUUM

SPARK ADVANCE FOR MAXIMUM POWER, CONTROLLED BY THROTTLE POSITION

SUPPLIED BY CENTRIFUGAL ONLY

IGNITION TIMING

	21	42
	18	36
	15	30
	12	24
	9	18
	6	12
	3	6
	0	0

CAR MPH	10	20	30	40	50	60	70	80	90	100
ENGINE RPM	500	1000	1500	2000	2500	3000	3500	4000	4500	5000
DISTRIBUTOR RPM	250	500	750	1000	1250	1500	1750	2000	2250	2500

SPARK ADVANCE

The centrifugal advance mechanism adjusts spark timing according to changes of engine speed, while the vacuum advance unit adjusts spark timing in relation to engine load. The two devices operate independently but the total spark advance is dependent on both.

An engine operating at 1500 rpm may be running under very light load with the throttle open only a small amount. This would be true if the car were being driven at a fixed speed on a level road. On the other hand, at the same speed, the engine may be under full load, as would be true if the car were being driven up a steep hill with the throttle wide open. In the first example, engine speed at 1500 rpm, engine under light load – the manifold vacuum would be high. This would operate the vacuum advance and the total spark advance would be the sum of the centrifugal advance plus the vacuum advance. In the second example, at the same engine speed but operating under full load, very little vacuum exists in the manifold. Because very little vacuum is available, the vacuum advance unit would return to its retarded position. Therefore, the total spark advance would depend on the operation of the centrifugal advance unit only.

Under any operating condition, the total spark advance is dependent upon the speed of the engine and on the load under which the engine operates. The higher the speed, the greater the advance, and also, the lighter the load, the greater the advance. At any speed, the load on the engine may vary from zero to full load depending on the position of the throttle plate. When the engine is operating at high speed, the throttle will be open which results in low manifold vacuum. Spark advance is controlled entirely by the centrifugal advance mechanism at this time.

In any event, the total spark advance at any engine speed is the initial advance, plus the centrifugal advance, plus the vacuum advance.

The dual diaphragm distributor provides both vacuum spark advance and vacuum spark retard action required to assist in the control of exhaust emissions.

EXHAUST EMISSION CONTROL SYSTEM ASSIST UNITS

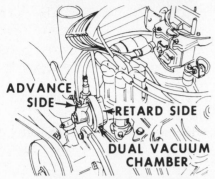

ADVANCE SIDE — RETARD SIDE

DUAL VACUUM CHAMBER

DUAL DIAPHRAGM SPARK CONTROL UNIT

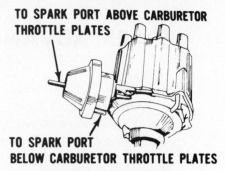

TO SPARK PORT ABOVE CARBURETOR THROTTLE PLATES

TO SPARK PORT BELOW CARBURETOR THROTTLE PLATES

DOUBLE-ACTING SPARK CONTROL UNIT

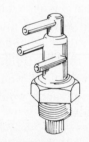

THERMOSTATIC VACUUM SWITCH

TO CARBURETOR VACUUM PORT ABOVE THROTTLE PLATES

TO CARBURETOR VACUUM PORT BELOW THROTTLE PLATES

TO FRONT OF DISTRIBUTOR VACUUM ADVANCE UNIT

DECELERATION VACUUM ADVANCE VALVE

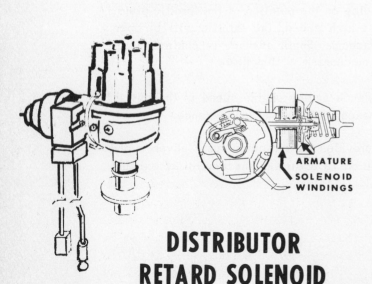

ARMATURE

SOLENOID WINDINGS

DISTRIBUTOR RETARD SOLENOID

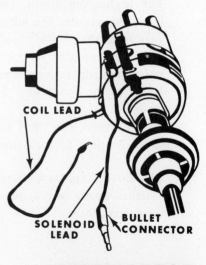

COIL LEAD

SOLENOID LEAD

BULLET CONNECTOR

DISTRIBUTOR ADVANCE SOLENOID

EXHAUST EMISSION CONTROL SYSTEM ASSIST UNITS

DUAL DIAPHRAGM VACUUM SPARK CONTROL UNIT

The Exhaust Emission Control System on some engines requires a retarded spark timing at idle speed and during periods of deceleration. The spark retard permits the use of a slightly greater throttle opening at idle to allow for increased air intake. This action provides more complete combustion during the idle and deceleration periods of engine operation which are the periods of greatest objectionable exhaust emissions.

The dual diaphragm distributor is used on some Ford Motor, General Motors and American Motors engines. The diaphragm housing contains two spring loaded independently operating diaphragms coupled by linkage to the movable distributor breaker plate. The forward diaphragm is moved by carburetor vacuum to shift the breaker plate against rotor rotation to advance the spark timing in the conventional manner. The rear diaphragm is actuated by intake manifold vacuum which moves the breaker plate with the direction of rotor rotation to retard the spark timing.

The centrifugal (mechanical) advance mechanism is, of course, not influenced by the vacuum double-acting control.

This dual-acting control of the vacuum diaphragm is made possible by what is called a "ported vacuum advance" arrangement. A vacuum line is connected into the carburetor at a point above the throttle plate(s). Vacuum at this point, with a closed throttle, is very weak. During periods of idle and deceleration intake manifold vacuum (which is high) is applied to the rear (retard) diaphragm thereby providing the desired degree of spark timing retard. As soon as the throttle is opened, the high carburetor venturi vacuum is applied to the front (advance) diaphragm and the spark timing is advanced in the conventional manner.

The dual diaphragm distributor actually affords three different phases of ignition timing. While the engine is being cranked for starting, there may be for example a 6° Before Top Dead Center setting. As soon as the engine starts idling, the vacuum retard may set the timing at 6° After Top Dead Center. As the vehicle is accelerated there will be varying degrees of spark timing advance, which will be a combination of the centrifugal and vacuum advance, depending on engine speed and load conditions.

DOUBLE-ACTING VACUUM SPARK CONTROL UNIT

Some Pontiac distributors used on exhaust emission controlled engines are equipped with a double-acting vacuum unit. This unit, however, contains only a single diaphragm but in addition to the usual vacuum line fitting at the front of the unit, there is another fitting mounted on the rear of the unit. The vacuum line connections are the same as those used in the Ford dual diaphragm unit. With this arrangement the single diaphragm unit is used to retard or advance the spark timing by applying a vacuum source to either the rear or the front chamber of the vacuum unit.

THERMOSTATIC VACUUM SWITCH

As previously stated, many exhaust emission control system equipped-engines idle with the ignition timing retarded. These engines also employ leaner calibrated carburetors. Because of the greater heat generated in an engine idling with retarded timing and lean fuel mixtures, there is a tendency for the engine to overheat during periods of prolonged idling, particularly during warm weather. To prevent this overheating possibility a thermostatic vacuum switch is employed. This valve is also called a distributor vacuum control valve or a ported vacuum switch.

There are three vacuum ports on the switch. They are usually identified by letters as, "D" for distributor, "C" for carburetor, and "M" or "MT" for manifold (intake manifold). Vacuum hoses connect the switch ports to their respective sources. On some installations the lower port is fitted with a filter to sense ambient pressure.

The switch is mounted in the engine block water jacketing or in the water distributing manifold where it can sense the temperature of the engine coolant.

When engine operating temperature is normal the thermostatic switch does not function and does not influence ignition timing. But in the event the engine coolant temperature rises above normal (approximately 220° F.), the switch automatically closes off the "C" port shutting off the ported carburetor (retard) vacuum to the distributor and opens the "M" port applying full manifold (advance) vacuum to the distributor vacuum unit. The manifold vacuum applied to the distributor results in spark advance and an increase in engine idle rpm with a corresponding increase in coolant circulation and fan action.

When the coolant temperature drops to normal value, the switch automatically shuts off the manifold vacuum to the distributor unit and reapplys the carburetor ported vacuum. The ignition timing now returns to its retarded setting and the idle rpm drops to its slow idle speed.

DECELERATION VACUUM ADVANCE VALVE

Another valve used in ignition spark control of some exhaust emission control equipped-engines is a Deceleration Vacuum Advance Valve. Some engines, may have a tendency toward a popping noise in the exhaust system during periods of deceleration "coast down" or during gear shifting. This condition is prompted by overretarding of the ignition timing.

To prevent this condition from occurring, the Deceleration Vacuum Advance Valve momentarily switches the vacuum to operate the distributor vacuum advance unit from its carburetor (low vacuum) source to a manifold (high vacuum) for just a few seconds and then back to its carburetor vacuum source.

The Deceleration Vacuum Advance Valve had popular application on 1966-67 Chrysler-built CAP-equipped engines and a somewhat lesser application on 1968-69 CAS-equipped engines. Some Pontiac and Ford models also employ this valve.

246

DISTRIBUTOR RETARD SOLENOID

An electric solenoid, built into and controlling the action of the distributor vacuum unit is used on some 1970-71 Chrysler-built V-8 engines. The function of the distributor solenoid is to provide spark timing retard during periods of idle and closed-throttle deceleration. This spark retard action assists in the control of oxide of nitrogen exhaust emissions.

The distributor solenoid is energized and de-energized by the carburetor throttle stop (solenoid) unit and curb idle adjusting screw. The distributor solenoid is equipped with two electrical leads. The inner (feed) lead is connected to the field circuit of the alternator and the outer lead is connected to ground at the carburetor throttle stop unit.

When the engine is idling, the curb idle adjusting screw contacts the carburetor throttle stop unit completing the circuit that energizes the distributor solenoid windings. The magnetic field produced by the windings attracts the armature against the solenoid core. The armature, being connected to the vacuum diaphragm link, shifts the distributor breaker plate in direction of cam rotation thereby retarding the spark. As soon as the engine is accelerated the idle adjusting screw breaks contact with the carburetor throttle stop unit and the distributor solenoid circuit is de-energized. Vacuum spark advance then occurs in the usual manner.

On engine restart, either hot or cold, the retard solenoid will not be functioning since the throttle will be opened slightly to start the engine. The separated throttle contacts interrupt the solenoid ground circuit and de-energize the solenoid. Vacuum spark advance is then applied in the usual manner.

DISTRIBUTOR ADVANCE SOLENOID
Some 1972 Chrysler-built V-8 engines have a distributor equipped with a timing advance solenoid. The function of the solenoid is to provide a 7½ degree spark advance in the ignition timing during the engine cranking operation. This action promotes better engine starting.

The solenoid is internally positioned in the vacuum unit. The only visible part is a terminal connector on the side of the distributor housing. The short lead attached to the connector terminates in a male bullet connector which also serves to identify the advance solenoid.

Power to energize the advance solenoid comes from the starting motor relay at the same connector that sends power to the starter solenoid. By using this type of connection the distributor advance solenoid is activated only while the engine is being cranked. Current to the solenoid energizes a coil that actuates a link which moves the distributor breaker plate against cam rotation to introduce a 7½ degree spark advance. As soon as the engine starts and the ignition key is released, the ignition timing returns to the basic timing setting.

TRANSMISSION CONTROLLED SPARK (TCS)

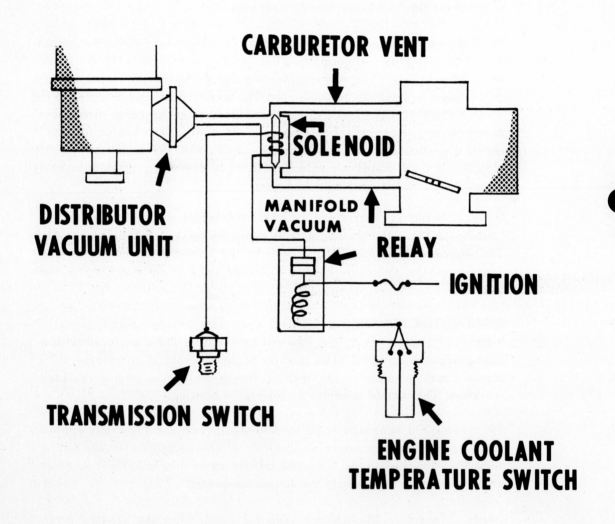

CARBURETOR VENT

SOLENOID

DISTRIBUTOR VACUUM UNIT

MANIFOLD VACUUM

RELAY

IGNITION

TRANSMISSION SWITCH

ENGINE COOLANT TEMPERATURE SWITCH

TRANSMISSION CONTROLLED SPARK

(TCS) SYSTEM BY GENERAL MOTORS

NO$_x$ SYSTEM BY CHRYSLER CORPORATION

Starting with their 1970 models, the five divisions of the General Motors Corporation introduced the system of Transmission Controlled Spark (TCS). This system is used in conjunction with their Controlled Combustion System (CCS) to essentially control the oxides of nitrogen portion of the exhaust emissions.

The TCS System is designed to prevent vacuum spark advance from occurring in first and second gear vehicle operation with either manual or automatic transmission. Full vacuum spark advance is permitted only in high gear and in reverse.

The TCS System employs a vacuum control solenoid fitted with either one or two vacuum lines. When one line is used, it is fitted to the carburetor below the throttle plate. When two lines are used, one line is fitted above the throttle plate and the other below the plate. Also connected to the solenoid vacuum valve are the electrical leads from the transmission control switch and from the engine coolant temperature switch.

In the first and second transmission gears the solenoid is energized by the transmission switch shutting off engine vacuum to the distributor vacuum unit thereby preventing vacuum spark advance. In high gear the transmission switch opens permitting the solenoid to de-energize and close the vent port. The vacuum port now opens permitting engine manifold vacuum to act on the distributor vacuum unit to advance the spark in the usual manner. On manual transmissions the switch is actuated by the gear shifter shaft position. On automatic transmissions the switch is operated by transmission fluid pressure.

There are two engine operational conditions that require an "override" of the TCS action on certain engines. One is cold engine drive-away and the other is during periods of engine overheating.

The cold override and the hot override switches are controlled by engine coolant temperature. When engine temperature is below 63° F. or over 232° F. the cold or the hot override switch de-energizes the TCS System and full vacuum spark advance is applied in the lower gear ranges to permit better cold-engine driveability or to reduce coolant temperature.

The TCS System is designed to be fail-safe. In the event any electrical control unit should fail, full engine manifold vacuum will be continuously applied to the distributor.

Any failure in the proper operation of the TCS System will result in either of two conditions. Continuous vacuum advance in the low gears will increase the exhaust emissions and prevent the vehicle from passing Federal emission standards. Lack of vacuum advance in high gear will result in loss of acceptable performance and an increase in fuel consumption.

When certain distributor tests are being conducted, as total degrees advance at 2000 or 2500 engine rpm, the vacuum unit must be in operation. By disconnecting the transmission lead connector at the solenoid (usually positioned next to the carburetor) the vacuum unit becomes operative and the distributor tests can be conducted.

THE NO$_x$ SYSTEM

The system of nitrous oxide emission control (NO$_x$) introduced on some 1971 Chrysler-built vehicles is basically similar in design and operation to General Motors Transmission Controlled Spark System. The essential differences being that the thermal switch is mounted on the firewall, rather than in the engine water jacketing, where it senses the ambient temperature in the plenum chamber. Some Chrysler engines, however, are also equipped with a coolant temperature switch, the same as used in the TCS System, to provide NO$_x$ system by-pass in the event of engine overheating. The other difference is that in the automatic transmission-equipped vehicles the speed control switch is mounted on the transmission housing inline with the speedometer cable to mechanically sense vehicle speed rather than being transmission oil pressure actuated as on the TCS System.

MAINTENANCE

No maintenance is required on any of the components of the TCS or NO$_x$ Systems. If any unit is malfunctioning, it must be replaced.

ELECTRONIC DISTRIBUTOR MODULATOR

ELECTRONIC DISTRIBUTOR MODULATOR

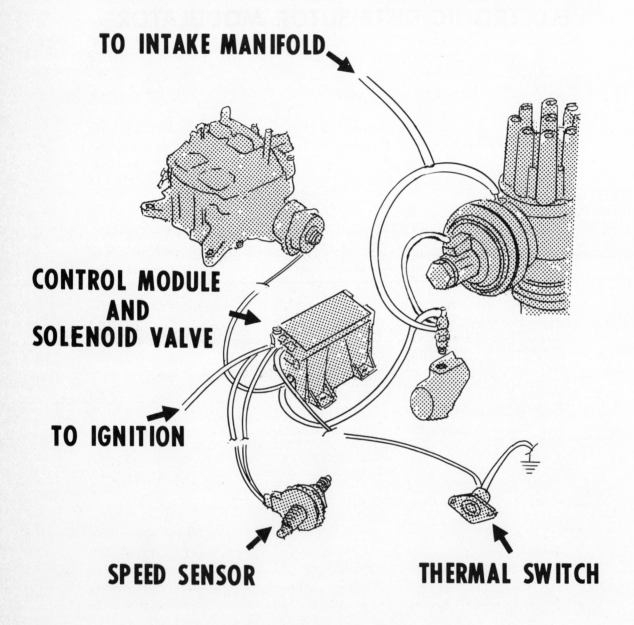

TO INTAKE MANIFOLD

CONTROL MODULE
AND
SOLENOID VALVE

TO IGNITION

SPEED SENSOR

THERMAL SWITCH

ELECTRONIC DISTRIBUTOR MODULATOR

Starting with the 1970 models, the Ford Motor Company equipped some of its engines with a new system of ignition spark control called the Electronic Distributor Modulator. The function of this system is to further assist in the control of objectionable exhaust emissions by increased vacuum spark control.

At vehicle speeds below 23 (± 2.3) mph on acceleration and below approximately 18 mph on deceleration, there is no vacuum spark advance. Previously the dual diaphragm system retarded the spark timing during periods of idle and deceleration only. With the addition of the Modulator System the spark is further controlled during periods of acceleration in the lower gears.

The Electronic Distributor Modulator System consists of four separate units which are added to the present distributor vacuum control system. The modulator components are: a speed sensor; a thermal switch; an electronic control module (or amplifier) which contains a three-way solenoid control valve.

The speed sensor is mounted between two sections of the speedometer cable and positioned just inside the driver's compartment. The sensor consists of a rotating magnet, driven by the speedometer cable, which turns inside a stationary field winding which is wound on a nylon bobbin. The bobbin is insulated from ground. As the speed magnet is rotated in the magnetic field it induces a voltage in the field which is in direct proportion to the speed at which it is driven. This voltage is impressed on the electronic control modulator.

The temperature sensitive thermal switch is mounted on the outside of the cowl near the front door hinge pillar on either side of the cowl depending on the vehicle model. The function of the thermal switch is to override (or cancel out) the signal from the speed sensor if the outside air temperature (not the underhood temperature) is below 58° F. The distributor is thereby permitted to operate through the standard vacuum advance control system for acceptable cold-engine performance.

The electronic control amplifier and the three-way solenoid valve are contained in a plastic housing which is mounted in the passenger compartment under the dash panel. The printed board circuit in the amplifier receives the input frequency from the speed sensor and the signal from the thermal switch. These signals to the amplifier determine whether or not the solenoid will be energized.

The system functions in the following manner. When the ignition switch is turned on, power is supplied to the amplifier. As the vehicle is put in motion, a voltage frequency is generated by the speed sensor as it is turned by the speedometer cable. When vehicle speed reaches 23 (± 2.3) mph, the voltage signal is strong enough to trigger the solenoid permitting the carburetor spark port vacuum to be applied to the distributor primary vacuum side to advance the spark timing. At this time the vehicle is in cruising speed range and the combined centrifugal and vacuum spark advance are applied to the distributor.

As the vehicle decelerates to approximately 18 mph, the reduced speed sensor signal de-activates the solenoid, the carburetor vacuum is shut off from the distributor vacuum unit, and the spark timing is retarded.

At any time the ambient air temperature is below 58° F. (the switch preset-temperature) the thermal switch will close keeping the solenoid activated, regardless of vehicle speed, permitting full carburetor vacuum to be applied to the distributor for maximum spark advance in the conventional manner

Pontiac's 1971 V-8 engines coupled to automatic transmissions are equipped with a system of vacuum spark control similar to the Distributor Modulator System. Vacuum spark advance is permitted only in third gear at speeds above 35 miles per hour by the speed sensor mounted between the sections of the two-piece speedometer cable. The V-8 engines with manual transmissions and the 6-cylinder engines use the General Motors TCS System.

ELECTRONIC SPARK CONTROL (SCS) SYSTEM

ELECTRONIC SPARK CONTROL (ECS) SYSTEM

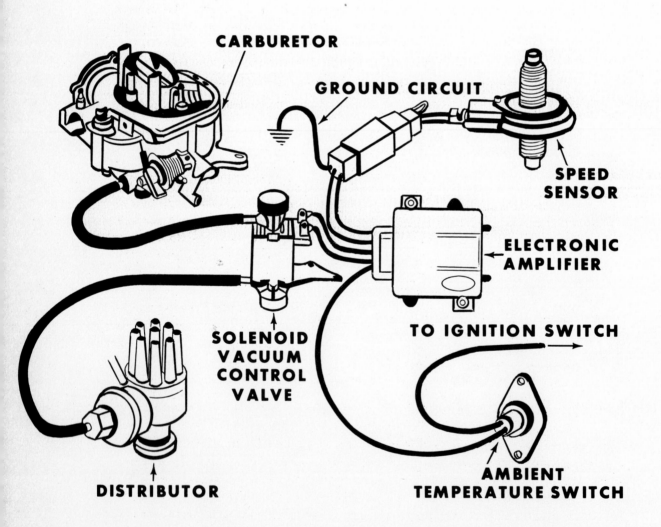

CARBURETOR

GROUND CIRCUIT

SPEED SENSOR

ELECTRONIC AMPLIFIER

TO IGNITION SWITCH

SOLENOID VACUUM CONTROL VALVE

DISTRIBUTOR

AMBIENT TEMPERATURE SWITCH

ELECTRONIC SPARK CONTROL (ESC) SYSTEM

The Electronic Spark Control System was introduced on some 1972 models of the Ford Motor Company, The function of the system is to prevent vacuum spark advance at lower vehicle speeds thereby inducing higher combustion temperatures that substantially reduce exhaust emissions.

The ESC System contains a speed sensor, an ambient air temperature switch, a solenoid vacuum control valve or distributor modulator valve and an electronic amplifier. A spark delay valve is used on some engines.

The electronic amplifier functions as a "switchboard" that controls the ECS System by the messages it receives from the speed sensor or the ambient air temperature switch. The amplifier is mounted on a printed circuit board located in the control module.

The speed sensor is located in either the engine or passenger compartment and is mounted in series with a two-piece speedometer cable. The sensor consists of a rotating magnet mounted inside a stationary winding. As the magnet is driven by the speedometer cable it generates a voltage in the field winding that is proportionate to the speed of the magnet or, in other words, to the speed of the vehicle. At speeds of 23, 28, 33 or 35 miles per hour, depending on the particular engine, the speed sensor generates a voltage signal that activates the control module and de-energizes the vacuum solenoid. Intake manifold vacuum is then applied to the distributor and vacuum spark advance is applied in the usual manner.

When a deceleration speed of approximately 18 miles per hour is reached, the electronic amplifier responds to the reduced voltage signal of the speed sensor, energizes the modulator valve and shuts off the vacuum supply to the distributor. At the same time the vacuum at the distributor vacuum unit is bled off through a vent at the bottom of the solenoid valve housing. Spark timing is then retarded.

If the engine coolant temperature should reach 230°F., the Ported Vacuum Switch overrides the control module to apply full manifold vacuum to the distributor advance unit regardless of vehicle speed. The increased idle speed accelerates coolant circulation and fan action restoring normal temperature at which time the module again controls the system.

The ambient air temperature switch overrides the control module to apply manifold vacuum to the distributor for full vacuum spark advance when outside air temperature is below approximately 49°F. to afford acceptable cold-engine driveaway and prevent a stall-on-start condition. At temperature is below approximately 65°F. the switch contacts close and the system functions by signals from the speed sensor. The temperature switch is located in a front door pillar where it is isolated from both passenger and engine compartment heat and can effectively sense outside air temperature.

Testing of the Ford Electronic Spark Control System should be performed by following the recommended Ford procedure and test sequence.

TRANSMISSION REGULATED SPARK (TRS) SYSTEM

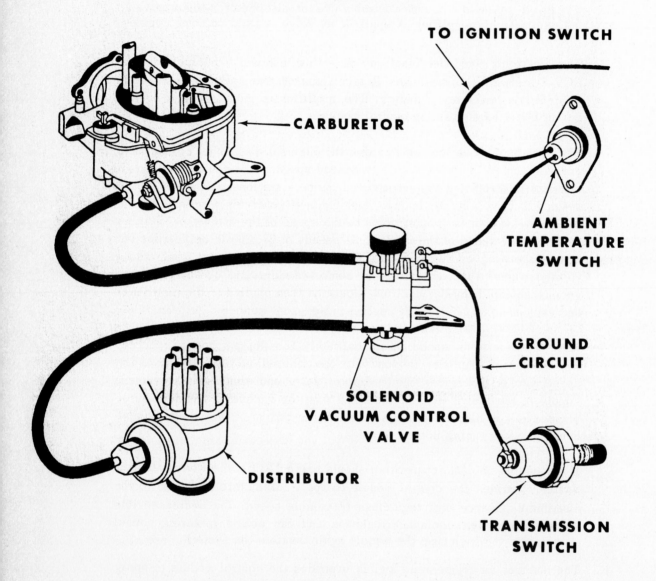

TO IGNITION SWITCH

CARBURETOR

AMBIENT TEMPERATURE SWITCH

GROUND CIRCUIT

SOLENOID VACUUM CONTROL VALVE

DISTRIBUTOR

TRANSMISSION SWITCH

TRANSMISSION REGULATED SPARK (TRS) SYSTEM

The Transmission Regulated Spark System is another emission control system designed by Ford Motor Company to limit the formation of oxide of nitrogen exhaust emissions. The rather simple TRS System resembles, to some degree, Ford Distributor Modulator and Electronic Spark Control Systems except that the electronic modulator and the speedometer-cable driven speed sensor are not employed.

The TRS System is controlled entirely by transmission gear selection. The system employs a transmission switch that serves as a circuit grounding device by sensing either automatic transmission high-gear hydraulic pressure or manual transmission shift linkage position; a solenoid-operated vacuum control valve; and an ambient air temperature switch. A spark delay valve may be used on some installations.

The system is designed to limit exhaust emissions by preventing vacuum spark advance in the lower gears. The fuel mixture and exhaust manifold temperatures are thereby increased by the retarded spark timing effecting more complete combustion with a reduction in hydrocarbon and carbon monoxide emissions. When the transmission switch senses high gear operation, the switch contacts are opened, de-energizing the vacuum control valve and opening the vacuum valve to allow manifold vacuum to be applied to the distributor advance unit. Vacuum spark advance is then applied in the usual manner.

If engine overheating occurs due to prolonged idling with retarded spark timing, the Ported Vacuum Switch overrides the solenoid vacuum switch and applies vacuum to the distributor regardless of transmission gear selection. The resulting idle speed increase assists in normalizing the coolant temperature.

At outside temperatures below approximately 49°F., the ambient air temperature switch contacts are open, de-energizing the TRS System, thereby overriding the solenoid vacuum switch. Vacuum is then applied to the distributor for vacuum spark advance regardless of transmission gear selection. The switch contacts close at temperatures above 65°F., energizing the system and returning control of the system to the transmission switch and solenoid vacuum switch. The temperature switch is positioned in a front door pillar where it can sense outside air temperature without being influenced by passenger or engine compartment heat.

Testing of the Transmission Regulated Spark System should be performed by following the test sequence recommended by Ford Motor Company.

SPARK DELAY VALVE

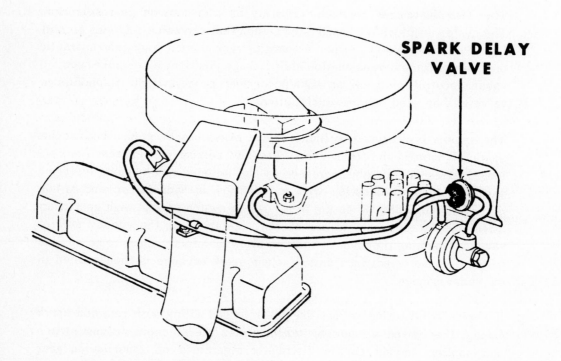

**SPARK DELAY
VALVE**

SPARK DELAY VALVE

The spark delay valve is another assist unit designed to control the ignition timing and limit the formation of exhaust emissions.

The valve is installed in the carburetor vacuum line at the distributor vacuum diaphragm on some Ford-built engines.

The function of the valve is to delay vacuum spark advance from occurring during rapid acceleration and to cut off the vacuum spark advance immediately on deceleration.

The delay valve cannot be tested or serviced. It must be replaced every 12,000 miles or 12 months, whichever occurs first. The length of the spark delay period varies with different engine applications. Several different valves are used and are color-coded for identification. It is important that the replacement valve be the same color as the original valve.

When installing a new valve, the black colored side of the valve must face the carburetor. The valve is designed for one-way operation and will not function if is installed backwards.

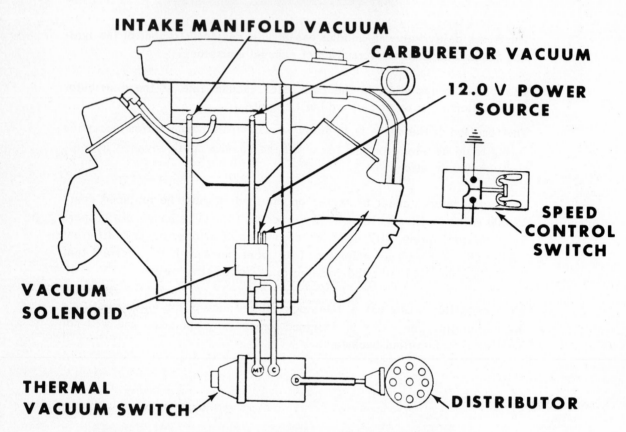

SPEED CONTROL SWITCH (SCS) SYSTEM

INTAKE MANIFOLD VACUUM

CARBURETOR VACUUM

12.0 V POWER SOURCE

SPEED CONTROL SWITCH

VACUUM SOLENOID

THERMAL VACUUM SWITCH

DISTRIBUTOR

MT   C   D

SYSTEM COMPONENT OPERATION

VEHICLE SPEED	SPEED CONTROL SWITCH	VACUUM SOLENOID
BELOW 33 $\pm$ 2 MPH, BELOW 25 MPH ON DECELERATION	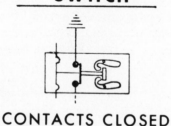 CONTACTS CLOSED	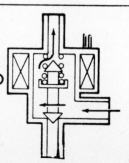 SOLENOID ENERGIZED, NO VACUUM ADVANCE
ABOVE 33 $\pm$ 2 MPH	 CONTACTS OPEN	SOLENOID DE-ENERGIZED NORMAL VACUUM ADVANCE

SPEED CONTROL SWITCH (SCS) SYSTEM

The Speed Control Switch System, introduced on the 1972 Cadillac, controls the distributor vacuum spark advance by vehicle speed rather than by transmission gear selection.

The SCS System is designed to limit exhaust emissions by shutting off the vacuum to the distributor at vehicle speeds below 35 miles per hour. A speed control switch sensor, which is the speed sensing device for the system, is mounted in the transmission, integral with the speedometer drive gear. The switch is designed to sense a vehicle speed of 33 ± 2 miles per hour. A vacuum solenoid positioned near the ignition coil controls carburetor vacuum to the distributor in response to the speed sensor signal.

At vehicle speeds of less than 33 ± 2 mph the vacuum solenoid is energized because the speed control switch contacts are closed providing a ground to complete the circuit. The carburetor vacuum supply to the distributor is shut off. The distributor vacuum unit is now vented through a hose at the rear of the carburetor to bleed all vacuum from the unit.

At speeds of 33 ± 2 mph, the speed switch contacts separate, breaking the circuit and de-energizing the vacuum solenoid. The pressure of its spring forces the solenoid plunger off the vacuum port seat and vacuum is applied to the distributor. More acceptable performance and increased fuel economy are effected by the vacuum spark advance at cruising speeds.

When vehicle speed is reduced below 25 miles per hour, the speed switch contacts close, completing the circuit which energizes the vacuum solenoid and returns it to the "no vacuum advance" position.

In the event a condition of engine overheating results from prolonged idling and low speed operation on retarded spark timing, the thermostatic vacuum switch will automatically apply full vacuum to the distributor vacuum unit regardless of vehicle speed, until normal operating temperature is restored.

COMBINATION EMISSION CONTROL (CEC) VALVE

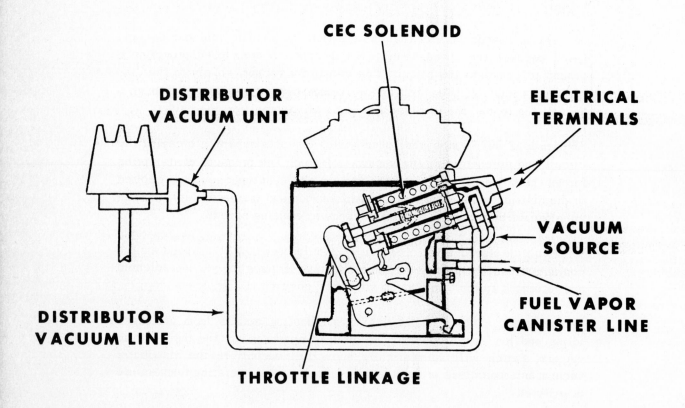

CEC SOLENOID

DISTRIBUTOR VACUUM UNIT

ELECTRICAL TERMINALS

VACUUM SOURCE

DISTRIBUTOR VACUUM LINE

FUEL VAPOR CANISTER LINE

THROTTLE LINKAGE

COMBINATION EMISSION CONTROL (CEC) VALVE

Chevrolet introduced the Combination Emission Control valve on its 1971 passenger cars and light trucks. The three-in-one, electro-magnetic CEC valve integrates the functions of three other emission control devices presently used by Chevrolet. These devices are: the Transmission Controlled Spark (TCS) solenoid, the carburetor idle stop solenoid, and the throttle slow-closing dashpot.

The CEC valve is about the size of a flashlight battery and is mounted as part of the carburetor assembly. It consists of a simple solenoid with a vacuum valve at one end and hexhead throttle check adjusting screw at the other end. The valve end of the solenoid is connected by a hose attached to a vacuum source and by a tube to the distributor vacuum advance unit.

When the ignition switch is turned ON, the valve is energized and is essentially "programed" thereafter by what the transmission does. Distributor vacuum spark advance is provided only in high gear and in reverse. In the lower gear ranges the valve is de-energized, vacuum to the distributor is shut off and the vacuum unit is vented to the atmosphere through a filter at the opposite end of the valve. A passage through the solenoid is provided by grooves moulded in the spool and by clearance between the adjusting screw and the plunger stop. When the transmission shifts into high gear the solenoid is energized, the vacuum port is uncovered and the plunger is seated in the opposite end of the solenoid shutting off the air vent. In this manner vacuum spark advance is applied only during high gear operation.

During periods of vehicle deceleration, the nonenergized CEC valve controls the throttle blade closing by providing a slow-closing action that permits as much air as possible to enter the carburetor for as long as possible, just short of interfering with normal braking, thereby reducing exhaust emissions.

As part of the system, three switches or relays are also employed. A thermostat coolant temperature switch energizes the CEC solenoid to provide a thermal override and permit vacuum spark advance when coolant temperature is below 82°F regardless of transmission gear selection. A time delay relay is incorporated to close the solenoid circuit for about 20 second after the ignition switch is turned ON. This action provided full vacuum spark advance to improve starting regardless of coolant temperature, although it only functions for a few moments. These two features improve vehicle driveability and eliminate the stall-after-start tendency. A reversing relay is provided to reverse the action of the transmission switch.

Some Chevrolet models equipped with air conditioning and automatic transmissions are idle speed adjusted with the air conditioning unit turned ON therefore requiring a higher idle speed. To avoid "Dieseling" on these models, a solid-state time device is employed that allows the air conditioning compressor clutch to become engaged for approximately 5 second after the ignition switch is turned OFF. The additional load the the compressor places on the engine effective reduces the "Dieseling" tendency.

Since the CEC valve controls throttle plate closing during deceleration, the curb idle speed can be somewhat reduced. The lower curb idle prevents "Dieseling" and reduces transmission creep eliminating the need for an idle stop solenoid, on most models. The slow throttle closing action also relieves the need for the conventional throttle slow-closing dashpot.

The CEC valve is factory preset to accommodate specific engine requirements and further adjustment is not required as part of a tune-up producere.

DISTRIBUTOR VACUUM CONTROL SWITCH

DISTRIBUTOR VACUUM CONTROL SWITCH

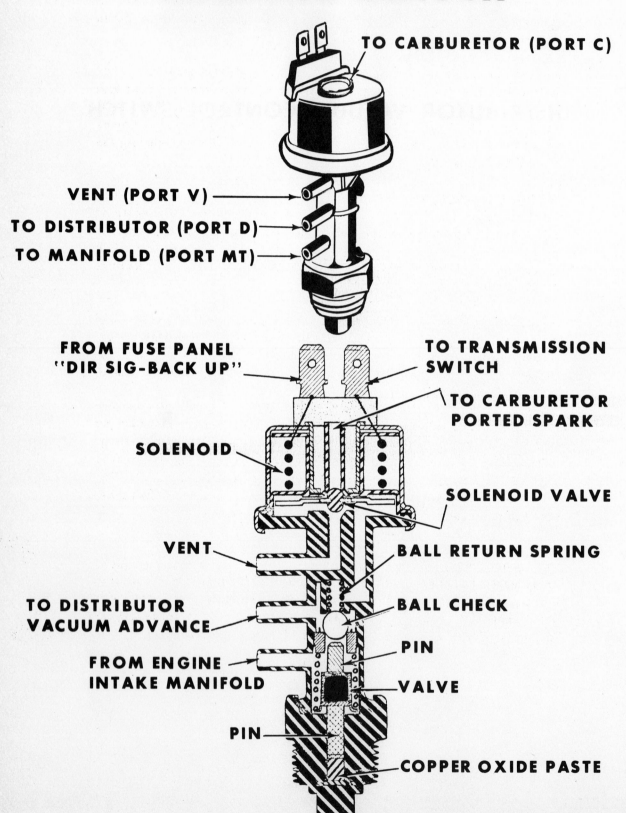

TO CARBURETOR (PORT C)

VENT (PORT V)

TO DISTRIBUTOR (PORT D)

TO MANIFOLD (PORT MT)

FROM FUSE PANEL "DIR SIG-BACK UP"

TO TRANSMISSION SWITCH

TO CARBURETOR PORTED SPARK

SOLENOID

SOLENOID VALVE

VENT

BALL RETURN SPRING

TO DISTRIBUTOR VACUUM ADVANCE

BALL CHECK

FROM ENGINE INTAKE MANIFOLD

PIN

VALVE

PIN

COPPER OXIDE PASTE

Chart No. 97

DISTRIBUTOR VACUUM CONTROL SWITCH

The Oldsmobile Distributor Vacuum Control Switch is another system designed to assist in the control of exhaust emissions. The switch is mounted in the water jacketing at the front of the engine where it can accurately sense the temperature of the engine coolant.

The function of the switch is to permit vacuum spark advance only when the transmission is in high gear or when an engine overheating condition exists as during periods of prolonged idle in hot weather. Since most emissions are formed in the transmission lower gear ranges, retarded spark timing while passing through the gears materially assists in reducing the formation of emissions.

The vacuum control unit is mounted on the upper portion of the switch assembly. When the transmission is in first or second gear, the solenoid is energized. The valve plunger seals off Port C from the carburetor and opens Port V, a vent to atmosphere that depletes any vacuum that might be present in the distributor advance unit. No vacuum spark advance occurs at this time.

During transmission high gear operation the solenoid is de-energized and the valve plunger moves down to seal off vent Port V. With this action, vacuum is directed from carburetor Port C and out distributor Port D to the distributor vacuum advance unit to provide vacuum spark timing advance.

In the event engine coolant temperature reaches 210°F., as during periods of prolonged idling in hot weather with retarded spark timing, the copper oxide paste positioned in the bottom of the switch expands and pushes the valve plunger upward sealing of vent Port V and opening intake manifold Port MT. Full manifold vacuum is then applied to distributor vacuum Port D and to the distributor vacuum unit. Vacuum spark advance is then introduced in the usual manner regardless of transmission mode.

The increase in idle speed caused by the vacuum spark advance results in increased coolant circulation and accelerated fan action which reduces the coolant temperature. When the coolant temperature drops, the oxide paste cools and contracts, permitting the ball check spring to reseat the ball and again restore vacuum spark control to the solenoid.

VACUUM SPARK CONTROL SYSTEMS

The formation of objectionable HC, CO and NO$_X$ exhaust emissions cocurs to a major extent during periods of idling and deceleration and in the transmission lower gear ranges. The popular method of controlling these emissions is by varying the vacuum supply to the distributor vacuum advance unit for more precise ignition timing control.

The spark control systems (Distributor Modulator, Electronic Spark Control (ESC), Transmission Regulated Spark (TRS), Speed Control Switch (SCS), Transmission Controlled Spark (TCS), Distributor Vacuum Control Switch and the Combination Emission Control (CEC) Valve) are all designed to apply the principle of varying vacuum control. They deny vacuum to the distributor at idle speed and while the transmission is passing through the lower gear ranges. The effect of the retarded spark timing insures optimum fuel combustion thereby limiting the formation of nitrogen oxides and hydrocarbons.

Vacuum spark advance will be introduced during periods of cold engine operation to afford acceptable cold-engine driveway and prevent the stall-on-start tendency, and during the periods of engine overheating to promote a faster idle speed to increase coolant circulation and accelerate fan action to normalize the coolant temperature on all systems.

The important point is that the only normal operating mode of the vehicle when the vacuum spark advance is applied, is when the transmission is in high gear or the vehicle is at cruising speed and the engine coolant is at normal operating temperature.

With this basic understanding in mind, some general thinking can be applied to **all** the systems.

- If any system fails to introduce vacuum spark advance when the engine is cold, or overheated, a temperature control unit such as an ambient temperature switch or a thermostatic vacuum control switch (or Ported Vacuum Switch), is malfunctioning.

- If the system has introduced vacuum spark advance at idle speed or in the transmission lower gear ranges (when there should be no spark advance), or if vacuum spark advance is not present during high gear operation (when there should be spark advance), a speed sensor, a transmission switch, a control module, or a solenoid is malfunctioning.

The above statement is somewhat simplified but the point is that the understanding of how one system functions pertains to a great extent to the other systems too. Thus the overall picture is not as complicated as it may at first appear.

270

IGNITION COIL CONSTRUCTION

IGNITION COIL CONSTRUCTION

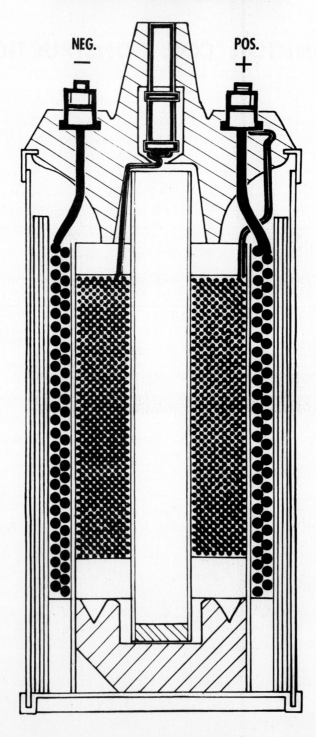

NEG. —   POS. +

IGNITION COIL CONSTRUCTION

An ignition coil is composed of a core, two windings, a metal case and a mounting bracket.

The core of the coil usually consists of thin soft iron strips or laminations. Its purpose is to increase the efficiency and output of the coil by promoting faster and more complete coil magnetic saturation. The soft iron core readily conducts magnetic lines of force, so less energy is used than if the magnetic lines of force had to travel through an air core.

The two windings are identified as a primary winding and a secondary winding. The primary winding consists of approximately 250 turns of relatively heavy wire, which is insulated with a special varnish. The secondary winding is wound inside the primary winding and consists of approximately 20,000 turns of very fine varnished wire. The many layers of the secondary windings are insulated from each other by high dielectric paper. One end of the secondary winding is connected to the high tension tower, while the other end is connected to one of the primary terminals inside the coil.

Ignition coils are often filled with oil or special compound to provide additional insulation and to help dissipate the heat that is created by the transformation of battery voltage. The dissipation of heat is very important in an ignition coil as heat tends to weaken insulation. An insulation breakdown results in partial or total coil failure.

IGNITION COIL ACTION

When the ignition switch is turned on and the breaker points are closed, current flows in the primary circuit. As current flows through the primary winding of the ignition coil, a strong magnetic field is produced, with the aid of the core. When the breaker points open, current ceases to flow through the primary windings of the coil and causes the magnetic field to collapse across the many thousands of turns of wire in the secondary winding. This action induces a very high voltage in the secondary circuit which forces current to jump the rotor and spark plug gaps.

When a piece of wire is connected across a source of voltage, current will immediately reach a maximum value determined by the resistance of the wire itself. But this is not true when the same wire is wound into a coil as in the primary winding of the ignition coil. This characteristic of a coiled conductor is called reactance or counterelectromotive force, and is due to the self-induced voltage in the coil.

When the breaker points close current starts to flow in the primary winding. As the magnetic field begins to build-up, the lines of force cut through the primary winding and induces a voltage that opposes battery voltage. Therefore, it takes a definite period of time for the primary current to reach a maximum rate of flow after the breaker points close. This period of time is called a "build-up" time. When maximum current is flowing in the coil winding, the maximum magnetic field is present and the coil is said to be fully "saturated".

If the breaker points remain closed for too short a period of time, maximum current flow will not be reached in the primary circuit and the maximum magnetic strength will not be attained. As a result, when the breaker points open, there will be less lines of force to cut through the secondary winding and coil output voltage will be reduced. This can cause the engine to misfire under certain operating conditions.

Reactance or counterelectromotive force not only opposes the build-up of current through the primary circuit, but also opposes any attempt to stop the flow of current. As the breaker points open, the magnetic field starts to collapse. The lines of force cut through the primary winding, but in the opposite direction from the build-up. This causes an induced voltage in the primary winding which is in the same direction as battery current and tends to keep current flowing. If current continues to flow when the breaker points open, there will be an arc between the breaker points. This arc has two very detrimental effects. First, it causes a transfer of metal from one point to the other resulting in point pitting. Second, unless the flow of primary current is stopped quickly, the magnetic field will collapse gradually and the secondary winding output voltage will be considerably reduced.

To control the arc that takes place between the points as they separate and to quickly stop the flow of primary current to develop maximum coil voltage, a condenser is connected across the points.

IGNITION COIL REPLACEMENT

As previously stated, the ratio of coil secondary turns to primary turns is approximately 100 to 1. Therefore a typical coil for a standard ignition system would have 200 turns of primary winding and 20,000 to 26,000 turns of secondary winding.

The coils used in transistorized ignition systems have a turns ratio of either 275 to 1 or 400 to 1. Because of the high current-carrying capacity of the heavier gauge transistor coil primary winding, approximately 95 turns of primary winding will be used in a coil having 26,000 turns of secondary winding.

It is very important when coil replacement is required the proper replacement coil be installed on the engine. Mixing standard and transistor system coils or coils of the wrong polarity will be quickly reflected in poor engine performance or total ignition failure.

274

CONDENSER CONSTRUCTION

CONDENSER CONSTRUCTION

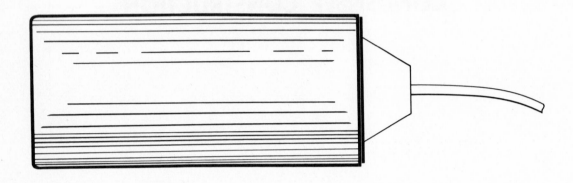

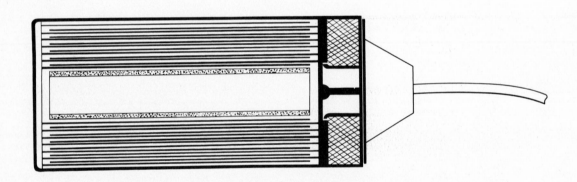

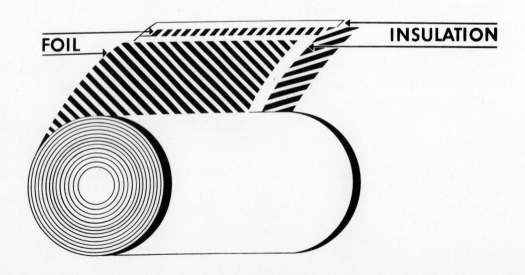

FOIL

INSULATION

CONDENSER CONSTRUCTION

The condenser is constructed of layers of aluminum foil, insulated from each other by layers of high dielectric insulating material. One layer of aluminum foil extends beyond the insulating material on one side, while the other layer of foil extends beyond the insulating material on the other side. The layers of aluminum foil and insulating material are then rolled into a tight cylinder and inserted into the condenser case. The layer of aluminum foil extending at one side will contact the bottom of the case and represents the ground terminal of the condenser. The other layer of foil will contact a disc which is connected to the insulated lead of the condenser.

Current does not flow through a good condenser. If it does, the two layers of foil are touching or there is a hole in the insulating material. If either condition exists the condenser is defective and must be replaced.

CONDENSER ACTION

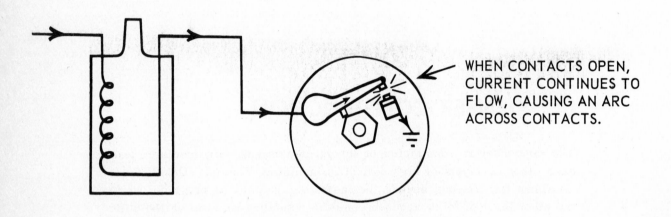

WHEN CONTACTS OPEN, CURRENT CONTINUES TO FLOW, CAUSING AN ARC ACROSS CONTACTS.

NO CONDENSER IN PRIMARY CIRCUIT

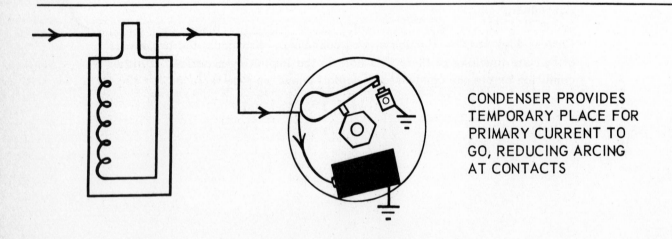

CONDENSER PROVIDES TEMPORARY PLACE FOR PRIMARY CURRENT TO GO, REDUCING ARCING AT CONTACTS

WITH CONDENSER IN PRIMARY CIRCUIT

CONDENSER ACTION

The functions of the ignition condenser are to reduce the amount of arcing across the points thereby preventing excessive metal transfer from one point to the other and to quickly stop the flow of current in the primary coil winding so that maximum voltage can be generated in the secondary coil winding.

The condenser can store a certain amount of electrical energy. When the breaker points open and the induced voltage in the coil tries to keep current flowing across the breaker points, the condenser will absorb the electrical energy until the breaker points have opened sufficiently so that an arc cannot occur. By the time the condenser becomes fully charged, the points have opened too far for arcing to take place.

By preventing the arc from occurring across the breaker points, the condenser brings the primary current flow to a sudden stop. This causes a very sudden collapse of the magnetic field, so that the lines of force cut through the windings in the coil with great speed. The voltage induced in the secondary winding forces current to jump the spark plug gap. The charge in the condenser surges back in a reverse direction through the primary circuit across the battery and builds up on the opposite plate of the condenser. The condenser then discharges in the opposite direction and charges the condenser once more in the original direction. Each time the condenser discharges, part of the energy is lost in overcoming the resistance of the circuit so that the oscillating current will die out, or nearly so, before the contact points close for the next build-up.

CONDENSER TESTING AND SELECTION

Condensers are tested for resistance, capacity and leakage. The resistance test reveals the presence of any loose or high resistance connections in the pigtail or the case. The capacity test checks the microfarad capacity of the condenser. The insulation test stresses the insulating material with about 500 volts while the tester meter indicates the presence of current leakage through the insulation. A condenser that tests defective in any test must be replaced.

The testing of a used condenser will reveal the presence of any defects but it cannot measure the amount of useful life left in the condenser. For this reason it is advisable to always install a new condenser when replacing the breaker points.

Some transfer of metal from one breaker point to the other is a normal action that occurs in all ignition systems. The function of a condenser of correct capacity is to prevent any excessive metal transfer. The capacity of the condenser is an important specification. It has been selected to match the particular requirements of a given ignition system. When replacing the condenser be sure the new condenser has the microfarad capacity recommended in the specifications. Typical specifications are: .18-.23 mfd; .21-.25 mfd; .25-.285 mfd.

When a good condenser of the correct capacity is employed and a condition of excessive point metal transfer nevertheless exists, the condition may be due to:

- Excessive primary or secondary circuit resistance in the ignition system which will upset the system balance.

- A high voltage regulator setting which causes increased primary current flow resulting in overheating and burning of the breaker points.

- Incorrect breaker point dwell angle.

- Continual high-speed operation.

- Frequent periods of prolonged idling.

Locating and servicing the cause of the trouble will prevent a reoccurrence of the condition. In the case of the last two conditions, more frequent tune-ups should be recommended.

DISTRIBUTOR CAP AND ROTOR

DISTRIBUTOR CAP
AND ROTOR CONSTRUCTION

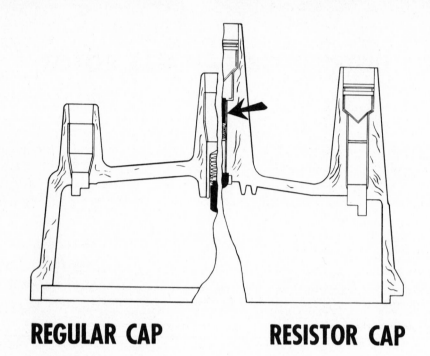

REGULAR CAP   **RESISTOR CAP**

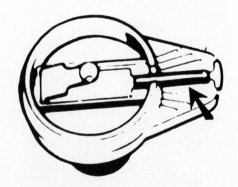

REGULAR ROTORS   **RESISTOR ROTOR**

DISTRIBUTOR CAP AND ROTOR

DISTRIBUTOR CAPS

A distributor cap is constructed of a high insulating bakelite material with metal inserts that are cast into the cap to receive the secondary wire from the coil to the distributor cap, and to receive the spark plug wires. These metal inserts extend downward inside the distributor cap so that the distributor rotor can provide a path between the center terminal and an outside terminal of the distributor cap.

A resistor-type cap containing built-in suppression usually includes a carbon resistor integral with the terminal in the cap coil tower. Due to the change characteristics of the carbon resistance over a period of time, it is advisable to change the distributor cap and rotor when performing a breaker point replacement.

DISTRIBUTOR ROTORS

Distributor rotors are usually constructed of materials similar to those used in distributor caps. A metal strip is used to form a conductor which contacts the center button of the distributor cap and carries the high tension current to the proximity of one of the secondary wire terminal posts inside of the cap. This metal strip does not actually close the circuit between the center button and the outer terminal. There is always an air gap between the end of the rotor tip and the distributor cap terminals. This gap is very small so that not more than 2000 to 3000 volts are required to carry secondary ignition current across this gap.

Some rotors incorporate a carbon resistor in their construction to provide ignition suppression. Due to the fact that carbon resistors occasionally undergo a change, unwanted resistance may be introduced by the loss of proper contact at the ends of the resistor. It is advisable, therefore, to change both the rotor and the distributor cap when replacing breaker points. Be sure to replace resistor-type rotors and caps with similar type units to retain the balance of the ignition system.

Rotors are keyed to the distributor shaft which places them correctly with relationship to the distributor cam. The relationship between the rotor and the distributor cam is critical, in that the rotor tip must be passing one of the secondary ignition wire terminals inside the cap at the time the ignition points open. To properly position the rotor on the centrifugal weight assembly of the General Motors V-8 engines and to prevent rotor breakage be sure to observe the shape of the locators on the underside of the rotor— one is round and one is square. They fit into similarly shaped receptacles in the weight base for proper rotor position.

SECONDARY CIRCUIT SUPPRESSION

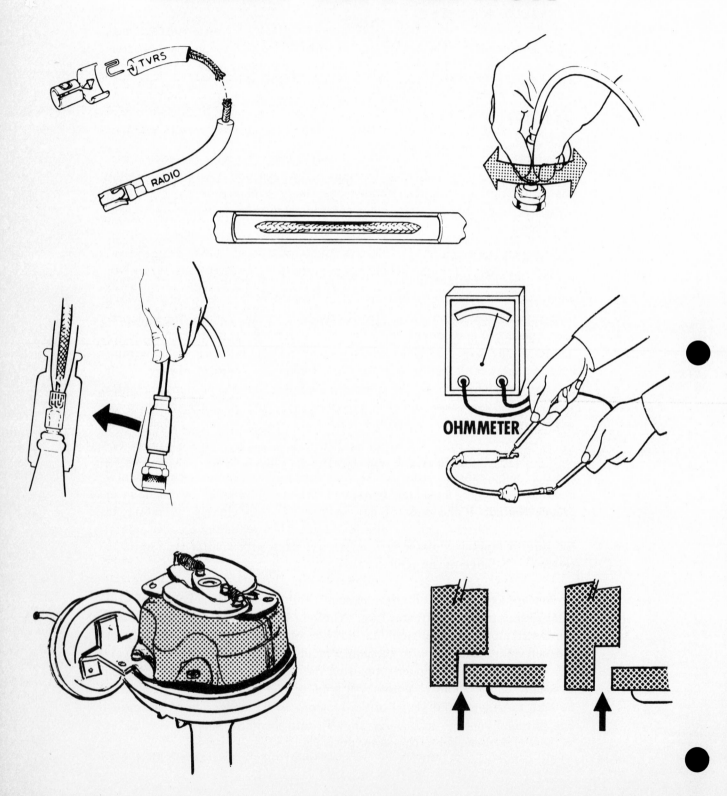

OHMMETER

SECONDARY CIRCUIT SUPPRESSION

Secondary circuit suppression is a purposely added resistance, and may be used in any part of the secondary system of the ignition circuit; in rotors, distributor caps, ignition wiring, or spark plugs.

Two desirable effects are claimed for secondary suppression. It reduces spark plug electrode wear or erosion by reducing the total amount of current flow across the spark plug gap. It also dampens high frequency radiation from the ignition system thereby reducing interference with radio and television reception.

Radio suppressors of the type used in the secondary ignition circuit are actually resistors. This added resistance placed in the secondary circuit does not reduce the voltage available at the spark plug gap. In order to lose voltage, current must be flowing. In the secondary ignition circuit, no current flows until the spark actually occurs at the spark plug gap. Therefore, until the spark plug fires, there is no voltage loss in the secondary circuit due to the resistance. Once the plug fires and current starts to flow, the voltage drop across the suppressor resistor reduces the total amperage that will flow across the spark plug gaps. This reduction in amperage tends to reduce the erosion or wearing away of the spark plug electrodes.

It was once believed that it was desirable to maintain current flow across a spark plug gap as long as possible to assure better ignition. However, if there is a combustible air-fuel mixture between the spark plug electrodes when the spark occurs, fuel burning will begin. Prolonging the duration of the spark will not add to the quality of ignition or the burning of the fuel.

RESISTOR IGNITION CABLES

In resistor spark plug cable design, the copper wire formerly used is replaced with a carbon or carbon impregnated linen core. These cables have a resistance value of approximately 4000 ohms per foot. Resistor cables introduced on the 1969 models of General Motors vehicles have a resistance value of about 2000 ohms per foot. Because of their design, resistor cables are particularly susceptible to damage by rough handling. These cables should never be "yanked" from the spark plugs. Instead, the rubber spark plug boot should be gently twisted to break the seal between the boot and the spark plug insulator. The cable should then be gently lifted straight from the plug.

Rough handling can break the connection between the cable and the plug terminal or it can stretch the cable sufficiently to change its resistance value. Any damage will be reflected in poor engine performance. Trouble of this nature may be difficult to locate since the symptoms are the same as spark plug malfunction. An ohmmeter is an excellent instrument with which to check cable continuity and to measure the resistance value of resistor cables.

Resistor cables are identified by the letters TVRS (Television Radio Suppression) or the word RADIO printed on the cable insulation.

RESISTOR SPARK PLUGS

Resistor spark plugs are designed with a 10,000 ohm resistor positioned in the center electrode. Resistor spark plugs designed for use with the 2000 ohm per foot resistance cables have a resistance value of 5000 ohms. As previously stated, the function of the resistor plug is to eliminate the "trailing end" of the spark at the plug electrodes after the plug has fired.

To retain the balance engineered into the ignition system, always replace resistor cables and spark plugs with the same type and resistance value originally used when performing your tune-up.

RADIO FREQUENCY INTERFERENCE SHIELD

Introduced on the 1970 GM V-8 models was the 2-piece Radio Frequency Interference Shield covering the distributor points and secured to the breaker plate by two screws. The function of the shield is to further assist in the reduction of TV and radio interference particularly on GM's windshield-mounted radio antenna installations. When replacing breaker points, be sure the point and condenser lead terminals are properly positioned, back-to-back, and are firmly secured to the connector. The terminals should then be bent slightly toward the cam be relieve the possibility of the terminals shorting out against the inside of the shield. Either slip-type or screw-type fastener breaker points may be used providing sufficient clearance exists between the point set connector and the inside surface of the shield. Before tightening the shield mounting screws be sure the primary ignition or condenser leads are not caught under the edge of the shield.

INCREASED ROTOR GAP

The increased rotor gap is another means employed to reduce TV and radio interference. The rotor gap prior to 1969 was approximately .025" and the increased gap is about .075". This design change makes it increasingly important to use the correct parts when replacing rotors and caps. Be sure to apply the increased KV (kilovolt) rotor gap specification to these installations if you are testing with an oscilloscope to avoid condemning good caps and rotors. The smaller gap KV specification was approximately 3 KV, the increased gap specification is approximately 8 KV.

COIL POLARITY

COIL POLARITY

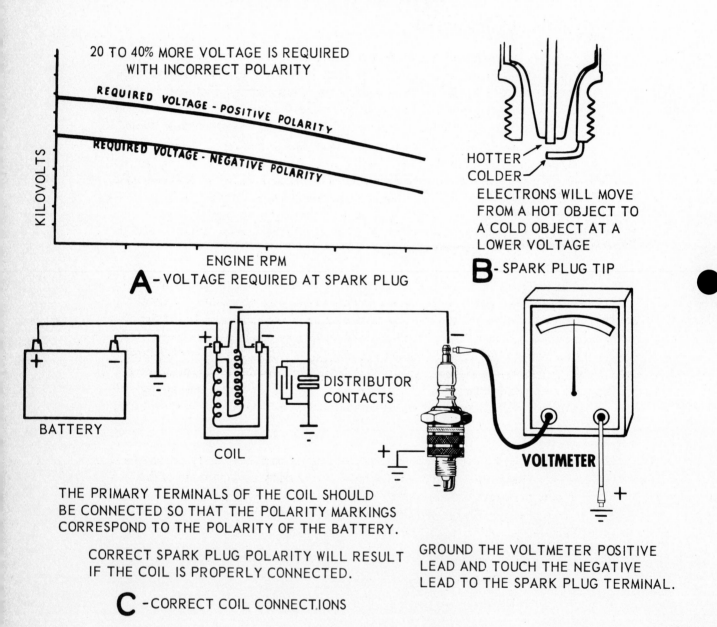

20 TO 40% MORE VOLTAGE IS REQUIRED WITH INCORRECT POLARITY

REQUIRED VOLTAGE - POSITIVE POLARITY

REQUIRED VOLTAGE - NEGATIVE POLARITY

KILOVOLTS

ENGINE RPM

A - VOLTAGE REQUIRED AT SPARK PLUG

HOTTER
COLDER
ELECTRONS WILL MOVE FROM A HOT OBJECT TO A COLD OBJECT AT A LOWER VOLTAGE

B - SPARK PLUG TIP

BATTERY

COIL

DISTRIBUTOR CONTACTS

VOLTMETER

THE PRIMARY TERMINALS OF THE COIL SHOULD BE CONNECTED SO THAT THE POLARITY MARKINGS CORRESPOND TO THE POLARITY OF THE BATTERY.

CORRECT SPARK PLUG POLARITY WILL RESULT IF THE COIL IS PROPERLY CONNECTED.

C - CORRECT COIL CONNECTIONS

GROUND THE VOLTMETER POSITIVE LEAD AND TOUCH THE NEGATIVE LEAD TO THE SPARK PLUG TERMINAL.

COIL POLARITY

To keep the required firing voltage of the ignition system as low as possible, the ignition coil must be connected for the correct polarity. The primary terminals of the coil should be connected so that the polarity markings correspond to the polarity of the battery, with the distributor connection considered as the ground. This will cause current flow through the spark plug from the center electrode to the ground electrode. This spark polarity, or secondary polarity, requires a lower voltage to fire the spark plugs since the electrons will be emitted from the hotter center electrode more easily than from the cooler ground electrode.

If secondary polarity is reversed, 20 to 40 percent more voltage is required to complete the secondary ignition circuit.

The polarity of the ignition coil can be checked with the aid of a voltmeter. Set the voltage range switch to the highest scale, connect the positive voltmeter clip to a good ground, and touch the negative voltmeter lead to any spark plug terminal. The meter needle should move upscale. You are not interested in a meter reading, only in the direction the needle moves. Should the needle move to the left or downscale, reverse polarity is indicated. This means the leads to the coil primary terminals are reversed, the battery is installed backwards, or the wrong coil is being used.

Ignition coils are not wound as negative or positive coils. Since ALL spark plugs have a positive ground, regardless of whether the vehicle's electrical system is negative ground or positive ground, it is advisable to make the spark plug polarity test every time service is performed on the ignition system or whenever the battery cables have been disconnected.

COMPARISON OF AVAILABLE AND REQUIRED SECONDARY VOLTAGE

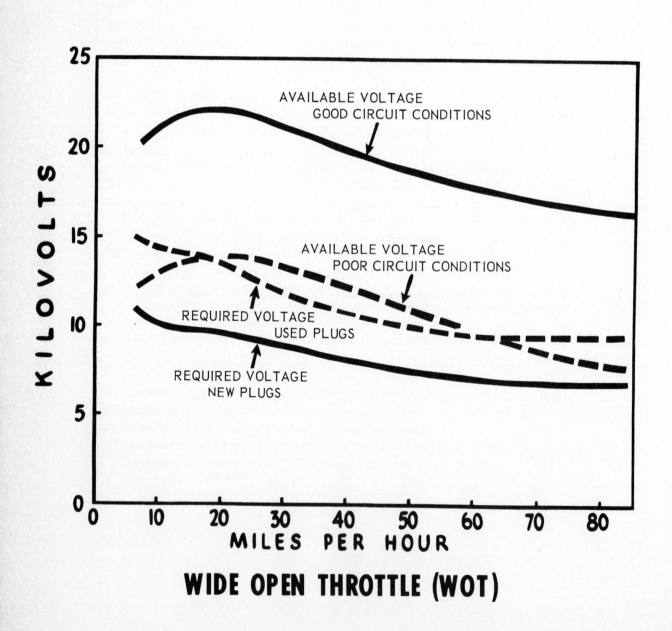

WIDE OPEN THROTTLE (WOT)

AVAILABLE VOLTAGE

Efficient ignition system performance requires that the available voltage, indicated by the upper solid line, is always higher than the required voltage, indicated by the lower solid line. If at any engine speed or load, the required voltage exceeds the available voltage, ignition failure results. This condition is indicated by the crossing of the dotted lines.

Any condition that reduces the safety margin between the available and required voltage, must be corrected before a tune-up can be considered successful. Reverse coil polarity, fouled spark plugs, plugs with eroded gaps, pitted breaker points, misaligned breaker points, improperly adjusted breaker points, leaking condenser, eroded distributor cap terminals, corroded distributor cap towers, defective rotor, loose spark plug cable terminals, damaged resistor cables or loose primary lead connections are all conditions that either introduce resistance into the ignition system thereby reducing the system's available voltage or increase the required voltage beyond the system's output capabilities.

CYLINDER NUMBERING SEQUENCES

IN-LINE ENGINES

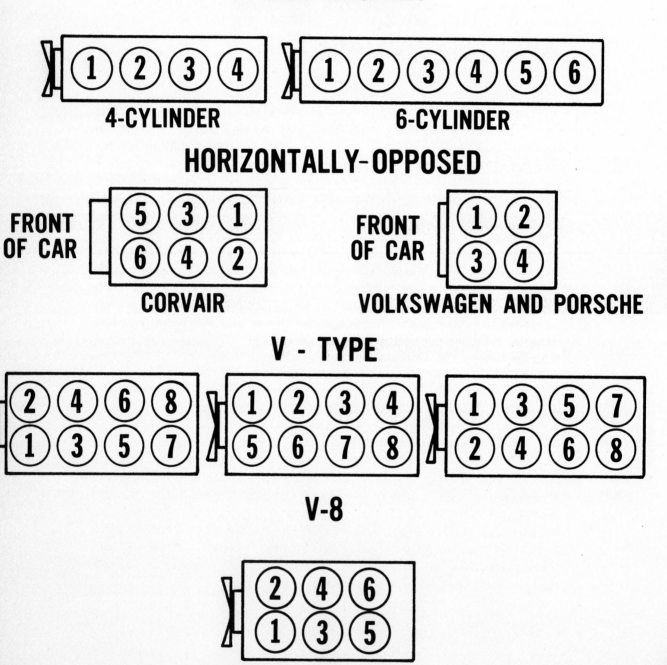

4-CYLINDER

6-CYLINDER

HORIZONTALLY-OPPOSED

FRONT OF CAR

CORVAIR

FRONT OF CAR

VOLKSWAGEN AND PORSCHE

V - TYPE

V-8

V-6

CYLINDER NUMBERING SEQUENCE AND FIRING ORDERS

Every engine has both a cylinder numbering and a firing order sequence. The cylinder numbers identify the cylinders according to their location in the engine. The cylinders are usually numbered from front to rear. The firing order of an engine is the listing of the cylinder numbers in the sequence in which they are fired.

There are three general arrangements of cylinders in automotive engines: in-line; V type; and horizontally opposed.

In the in-line design the cylinders are positioned one behind the other. This cylinder arrangement is typical of most 4-cylinder and 6-cylinder engines.

The V-type engine, popular in V-8 engine design, have 4 cylinders set in each of two banks with the banks set 90 degrees apart. A V-6 engine employs the V-type design with 3 cylinders in each bank.

The horizontally-opposed type engine design has two engine cylinder banks set 180 degrees apart giving the engine a flat appearance. The Chevrolet Corvair and the German Volkswagen and Porsche imports use engines of this design.

Regardless of the design of an engine or the number of cylinders it contains, the pistons in the engine move in pairs. That is, there are always two pistons which are attached to crankpins having a common centerline. The engine is designed in this manner to provide the necessary mechanical balance and because all the cylinders in an engine must be fired in two revolutions (720 degrees) of the crankshaft.

The location of No. 1 cylinder is important to the tune-up specialist because it is a reference point for ignition timing and for valve timing, if he is engaged in engine work. The No. 1 cylinder is usually the cylinder nearest the radiator. In the case of V-type engines, however, the No. 1 cylinder may be in either bank. The engine diagram chart illustrates the No. 1 cylinder position on popular production engines. Refer to your tune-up specifications for both the cylinder numbering sequence and the firing order of the engine you are tuning.

SPARK PLUG HEAT RANGE

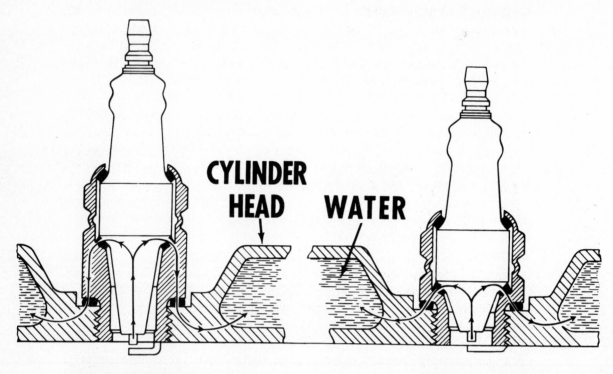

CYLINDER HEAD   WATER

HOT PLUG   COLD PLUG

SPARK PLUG REACH

PROPER REACH   REACH TOO SHORT   REACH TOO LONG

SPARK PLUG HEAT RANGE

To provide proper engine performance, spark plugs must operate within a certain temperature range. If the spark plugs operate at too cold a temperature, less than 700°F., soot and carbon will deposit on the insulator tips which cause fouling and missing. If the plugs run too hot, more than 1700°F., the insulator will be damaged and electrodes will burn away rapidly. In extreme conditions, hot plugs cause premature burning (preignition) of the air-fuel mixture.

The ability of a spark plug to transfer heat from the insulated center electrode tip is controlled by the design of the spark plug. The only path for heat to escape is through the insulator tip, spark plug shell and gasket, through the cylinder head, to the cooling liquid in the water jacket. By varying the length and shape of the insulator and shell, the manufacturer is able to produce spark plugs with different heat range characteristics and thereby control their operating temperatures.

A visual inspection of the spark plugs after they are removed from the engine will reveal the existence of a variety of engine ailments in addition to indicating, with reasonable accuracy, the correctness of the heat range of the spark plugs used. The nature of the deposits collected on the plug insulator and the condition and the gap of the electrodes also provide valuable clues. It is important, however, not to idle a cold engine for any length of time before removing the spark plugs for examination. The plugs taken from a cold engine that has been idling with partial choke will very likely be soot-fouled by the rich fuel mixture. This deposit will give a false impression of the true condition of the plug.

When replacing spark plugs during your tune-up, it is usually advisable to replace the plugs with the same heat range used as original equipment. If, however, the plugs in an engine constantly exhibit electrode wear and blistered insulators, a colder range of plugs should be used. If, because of constant low-speed city driving the plugs are constantly fouled, a hotter range of plugs should be installed. When a change is made, change only one heat range number at a time.

Remember — a hotter spark plug has a higher temperature, not a hotter spark.

SPARK PLUG REACH
Spark plug reach is the length of the threaded portion of the plug. If the reach is too short, the plug electrodes will be in a pocket and may misfire under certain conditions. The exposed threads in the cylinder head will "carbon up" making cleaning of the threads necessary before plugs of the proper reach can be correctly installed and torqued. If the reach is too long, the plug threads will be exposed and may overheat resulting in preignition. The exposed threads will also carbonize making plug removal difficult. The danger of piston head damage also exists if the plug reach is too long.

SPARK PLUG FEATURES

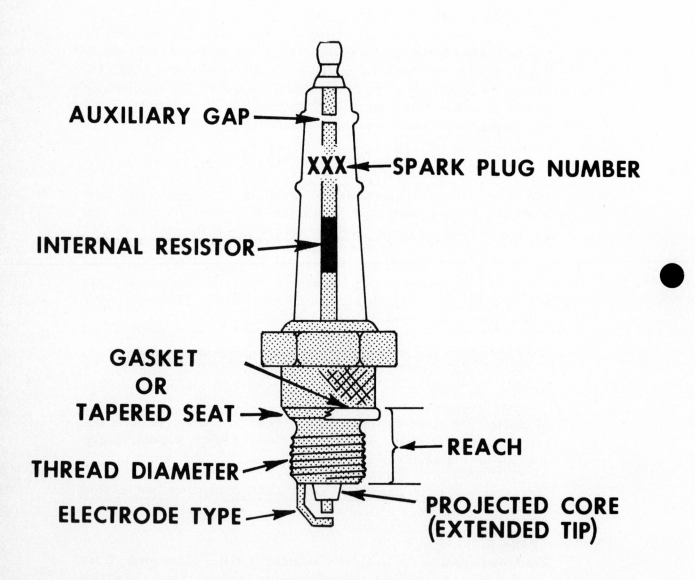

AUXILIARY GAP

XXX ← SPARK PLUG NUMBER

INTERNAL RESISTOR

GASKET
OR
TAPERED SEAT →

THREAD DIAMETER

ELECTRODE TYPE →

← REACH

PROJECTED CORE
(EXTENDED TIP)

SPARK PLUG FEATURES

Besides the important Heat Range and Reach requirements of the spark plug, there are several other features that are highly instrumental in proper spark plug performance. Since every feature is important in proper plug selection, the tune-up specialist should understand what the plug markings represent.

To the basic requirements of heat range and reach are special design features, as: tapered seat; internal series booster gap; internal resistor; two ground electrodes; projected firing tip; heavy-duty electrodes and other features.

To assist the tune-up specialist in the proper selection of spark plugs for the engine he is tuning, the plug manufacturers mark their plugs with numbers and letters to identify every specification and every feature. The following are only a few popular examples of these markings.

CHAMPION
Champion spark plugs markings have the following meanings:
- The prefix (first) letter indicates the thread diameter and plug reach. Examples of popular application are: J is 14mm thread, 3/8" reach; L is 14mm thread, 1/2" reach; N is 14mm thread, 3/4" reach; and F is 18mm thread with tapered seat.
- If the plug has any special design features, a prefix letter indicating the feature will be placed before the first prefix letter. Example: B is a tapered seat; U is a booster gap; and X is a built-in resistor.
- The number following the prefix letter(s) is the heat range. If the range of numbers are, for example 3 to 18, then Number 3 plug is the coldest and the number 18 plug is the hottest. In between these two numbers is a gradual increase in heat range with every increase in number.
- The suffix (last) letter indicates the spark gap design. Examples: B is a two ground electrode; Y is a projected core nose.

Using the above data the following are a few popular Champion spark plug numbers.
- UJ-12Y. This plug has a booster gap (U); has a 14mm thread with a 3/8" reach (J); is No. 12 (12) in the heat range; and has a projected core nose (Y).
- UBL-13Y. This plug has a booster gap (U); has a tapered seat (B); has a 14mm thread with a 1/2" reach (L); is No. 13 (13) in the heat range and has projected core nose (Y).

- XJ-11. This plug has a built-in resistor (X); has a 14mm thread with a 3/8" reach (J); and is No. 11 in the heat range.

AC
The spark plugs manufacturered by the AC Division of General Motors Corporation are identified in a somewhat similar manner:
- The prefix (first) letter indicates the spark plug features. Example: B is a series gap; C is commercial (heavy-duty) electrodes; R is a built-in resistor.

- Following the prefix letter are two numbers. The first of the two numbers indicates the thread size. All numbers starting with 2 are 1/2", all numbers starting with 4 are 14mm; all numbers starting with 7 are 7/8", among others. The second number indicates the heat range. The higher the number the hotter the plug. A 46 plug (6) is hotter than a 44 plug (4) although both of the plugs have a 14mm thread by virture of the first No. 4.
- Suffix letters following the numbers indicate special design features. Examples: XL is extra long reach; S is extended tip; T is tapered seat; and TS is tapered seat with extended tip.

Following are a few AC spark plug numbers using the above data:
- 45S. This plug has a 14mm (4) thread; is No. 5 in the heat range (5); and has extended tip (S).
- 44TS. This plug has a 14mm thread (4); is No. 4 in the heat range (4); and has a tapered seat with an extended tip (TS).
- 45XL. This plug has a 14mm thread (4); is No. 5 in the heat range (5); and has an extra long 3/4" reach (XL).

AUTOLITE
On Autolite spark plugs the first letter is the thread size. Examples: A is a 14mm; B is an 18mm.
- The second letter is the reach. Examples: None is 3/8"; L is 7/16"; E is 1/2"; and G is 3/4".
- If there are other second or third letters they represent other special design features. Examples: R is a built-in resistor; F is a tapered seat; and Z is an internal series gap.
- Following the prefix letter(s) are numbers that represent the heat range. The numbers run from 2 which is the coldest plug to 11 which is the hottest. Autolite classifies their heat range numbers into three catagories. No.'s 2 through 5 are cold plugs; No.'s 6 through 8 are medium plugs; and No.'s 9 through 11 are the hot plugs.
- Letters after the heat range numbers are special features. Example: S is for Shield; X is for Outboard Marine; and M is for Moisture-Proof Pack.

A few typical examples of Autolite spark plugs using the above symbols are:
- A5. This plug has a 14mm thread (A); has a 3/8" reach (because of no number); and is No. 5 in the heat range (cold).
- AE6. This plug has 14mm thread (A); has a 1/2" reach (E); and is No. 6 in the heat range (medium).
- AER6. This plug has a 14mm thread (A); has a 1/2" reach (E); has an internal resistor (R); and is No. 6 in the heat range (medium).

It can be readily seen from the few examples cited that there is a lot more to the proper selection of a set of spark plugs than merely "picking a number." At best the wrong plugs will result in poor engine performance and short plug life - and at worst, they can cause severe engine damage for which the mechanic will be held responsible.

298

TRANSISTOR IGNITION SYSTEM

BREAKER POINT TYPE

TRANSISTOR IGNITION SYSTEMS
BREAKER POINT TYPE

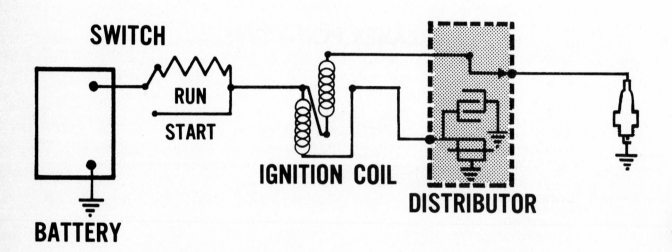

CONVENTIONAL IGNITION SYSTEM

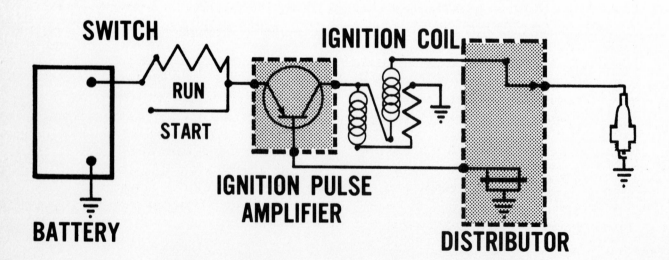

TRANSISTOR IGNITION SYSTEM

TRANSISTOR IGNITION SYSTEM

BREAKER POINT TYPE

The transistorized ignition system has been developed because the increased power outputs and higher sustained speeds of the modern high compression engine have taxed the conventional ignition system to its capacity.

The conventional ignition system has one limiting design factor that puts a ceiling on the system's capabilities. This factor is the approximate 2 ampere current flow capacity of the breaker points. If the points could handle about 6 amperes without excessive point pitting, the system would have a substantially greater high speed output. Besides this basic shortcoming of the standard ignition system, is the fact that during high speed operation the breaker point dwell angle is of such short duration that complete coil magnetic saturation is impossible. The transistorized ignition system overcomes both these undesirable factors.

The top illustration on the chart is a schematic of the conventional ignition system. The breaker points are in series with the ignition coil primary winding. The current carrying capacity of the points is the controlling agent that decides how much current can pass through the coil. With a limited strength in the primary circuit there is a proportionately limited secondary circuit output.

The lower illustration on the chart is a elementary schematic of a transistorized ignition system that employs breaker points to mechanically "trigger" the transistor base circuit. The breaker points are now connected to the base circuit of the transistor. As we discussed in our coverage of function, construction and operation of the transistor, the breaker points in the transistor ignition system are used to merely "trigger" the base circuit. In this manner the emitter-collector circuit is turned ON and OFF.

Since the base circuit has a very light current flow, the breaker points now handle only about ½ ampere. Arcing across the points is practically nonexistent with this light current flow so the conventional condenser has been eliminated. A current flow of approximately 6 amperes can be used in the coil primary circuit. Coil magnetic saturation is now virtually instantaneous with a proportionate increase in secondary circuit output. With a coil designed to accommodate the greater primary current flow, coil saturation is definitely faster than in the conventional ignition system and virtually unaffected by high speed operation.

The benefits of the transistorized ignition system can be summed up by these facts:
1. A small current flow across the breaker points,
2. can be used control a much greater current flow through the coil primary windings,
3. to produce a higher and more persistent secondary voltage,
4. that is virtually unaffected by high speed operation.

The complete transistor ignition circuit also includes condensers, a Zener diode, resistors and capacitors. Most of these units are mounted in an amplifier assembly which also acts as a heat sink to draw off excessive heat.

When a transistorized ignition system is shop installed on a vehicle, it is a rather common practice to leave the conventional ignition system intact. Then, through the aid of a switch or by the transfer of wires, either system is available in the event of failure of the other.

TRANSISTOR IGNITION SYSTEM

MAGNETIC PULSE TYPE

TRANSISTOR IGNITION SYSTEMS

MAGNETIC PULSE TYPE

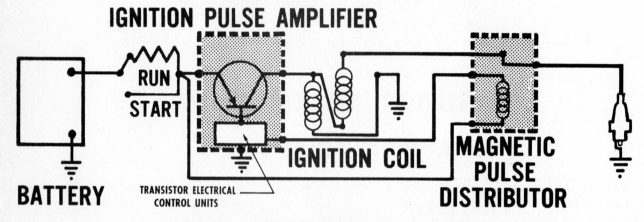

IGNITION PULSE AMPLIFIER

RUN

START

BATTERY

TRANSISTOR ELECTRICAL
CONTROL UNITS

IGNITION COIL

MAGNETIC
PULSE
DISTRIBUTOR

TRANSISTOR IGNITION SYSTEM

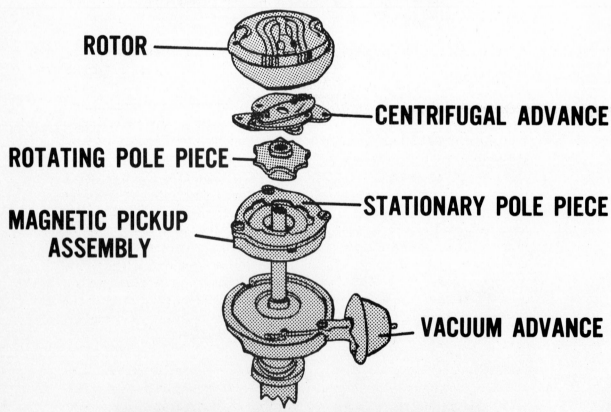

ROTOR

CENTRIFUGAL ADVANCE

ROTATING POLE PIECE

STATIONARY POLE PIECE

MAGNETIC PICKUP
ASSEMBLY

VACUUM ADVANCE

MAGNETIC PULSE DISTRIBUTOR

TRANSISTOR IGNITION SYSTEM

MAGNETIC PULSE TYPE

The magnetic pulse type transistor ignition system electrically "triggers" the base circuit of the transistors in the amplifier.

The breaker plate and point set of the conventional ignition system is replaced by a magnetic pickup assembly which consists of a permanent magnet with an attached stationary pole piece and a pickup coil. The stationary pole piece mounted on the permanent magnet has equally spaced internal teeth. There is one tooth for each cylinder in the engine. The standard breaker cam is replaced with a rotating pole piece. This pole piece has equally spaced external teeth or vanes, one vane for each cylinder. Movement of the pickup assembly is controlled by the standard vacuum advance unit. The rotating pole piece is moved by the conventional centrifugal advance weights. By these two methods the spark advance of the magnetic pulse transistor distributor is controlled in the conventional manner.

The ignition pulse amplifier contains three transistors, resistors, diodes, capacitors and wiring or a printed circuit panelboard. There are no moving parts in the amplifier. The Zener diode serves to protect the transistors from high voltages induced in the primary circuit.

The magnetic pulse transistor ignition system functions as follows. As the distributor shaft turns, the vanes of the rotating pole piece are not aligned with the internal teeth of the magnetic pickup assembly pole piece. At this time current will flow through two "switching" transistors in the amplifier, through the coil primary winding and to ground permitting a strong magnetic field to saturate the coil. As the distributor shaft turns further, the vanes of the rotating pole piece are aligned with the internal teeth of the magnetic pole piece. At this instant a magnetic path is established through the center of the pickup coil inducing a voltage in the coil. This voltage causes a current flow in the third transistor in the amplifier, the "triggering" transistor, to turn off the other two transistors. Current flow in the coil primary winding stops, the magnetic field collapses and high voltage is induced in the coil secondary winding to fire the spark plug. As the distributor shaft continues to rotate, the misalignment of the teeth on the rotating pole piece with the internal teeth of the magnetic pole piece again permits the "switching" transistors to complete the coil primary circuit and the entire action is repeated.

The transistor ignition system coil has fewer and heavier primary turns and a higher turns ratio of primary and secondary windings than a conventional ignition system coil. These features produce the desired maximum high voltage output through the high-speed ranges.

CAPACITOR DISCHARGE IGNITION SYSTEM

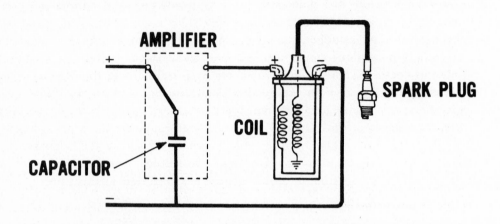

CAPACITOR BEING CHARGED

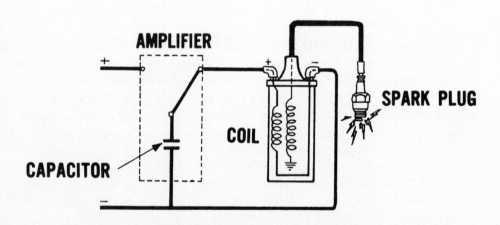

CAPACITOR DISCHARGING INTO COIL PRIMARY WINDINGS

CAPACITOR DISCHARGE IGNITION SYSTEM

The Capacitor Discharge (CD) ignition system has several units in common with the Magnetic Pulse-Type transistor ignition system - a special ignition coil, a transistorized pulse amplifier, and a magnetic pulse-type (breakerless) distributor.

The major difference of the CD ignition system is a high-voltage capacitor (condenser) connected across the coil primary windings. During the time the spark plugs are not firing (normal dwell time), the capacitor is charged by the transistorized amplifier with approximately 300 volts. A Zener diode (a voltage protection diode) in the amplifier, limits the circuit voltage to a 300 volt ceiling. On a voltage impulse signal from the magnetic pulse distributor, the capacitor discharges its high voltage charge into the coil primary windings. The step-up transformer action of the ignition coil increases this primary voltage to as much as 30,000 volts on the demand of the ignition system. By maintaining maximum primary circuit output, complete coil saturation and maximum secondary circuit output voltage is assured.

The Capacitor Discharge ignition system is said to possess several advantages over other ignition systems. Varying battery voltage due to charge and discharge cycles or lack of sufficient charging system output (when full electrical system demands are required) which occur on other ignition systems, reduce the primary circuit voltage. This, of course, results in a proportionate loss in secondary voltage output.

In the Capacitor Discharge system, if the battery has only a sufficient charge to turn the engine over, the amplifier will load the capacitor with a full charge of 300 volts. Consequently the ignition system will not be starved out due to voltage variations or voltage drop during starting. Maximum voltage is always available for even severe cold-weather starts.

Another desirable factor is that the high voltage spark is delivered in less than normal time. High voltage leakage, that occurs over distributor cap surfaces, moisture covered spark plug cables and fouled or gasoline-wetted spark plugs, is thereby minimized. The combination of consistent high-voltage impulses and fast spark delivery is said to make the cold weather starting of even a flooded engine much easier than with any other ignition system.

Besides having the remarkable engine starting capability, the constant primary circuit input to the coil assures maximum voltage output through the vehicle's entire speed range. There will be no "running out of ignition" at high speed. And the fact that the distributor is breakerless there will be no voltage fall-off due to high-speed breaker point bounce. Spark plugs with wider than normal gaps (resulting from high mileage) will also be consistently fired by the CD system.

The external identifying feature of the Capacitor Discharge Ignition system is the distinctive red color of the distributor cap and coil.

The mechanical switch indicated on the chart is for the purpose of explanation only. The actual charging of the capacitor and the switching from system charge-to-discharge is accomplished electronically by transistors, diodes, resistors, capacitors, a transformer and other units in the amplifier.

ELECTRONIC IGNITION SYSTEM

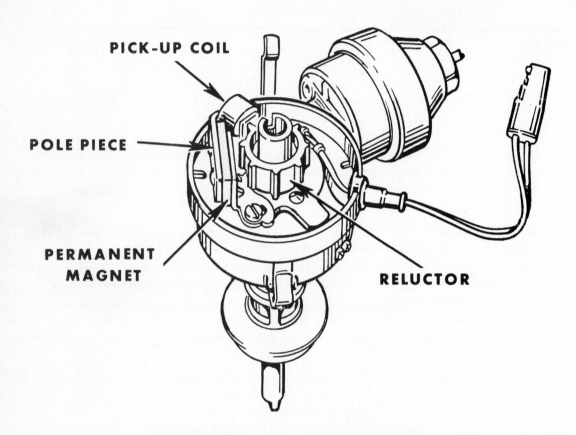

PICK-UP COIL

POLE PIECE

PERMANENT MAGNET

RELUCTOR

DISTRIBUTOR COMPONENTS

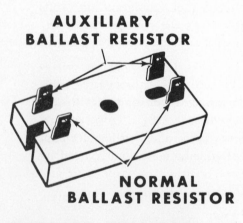

AUXILIARY BALLAST RESISTOR

NORMAL BALLAST RESISTOR

BALLAST RESISTOR

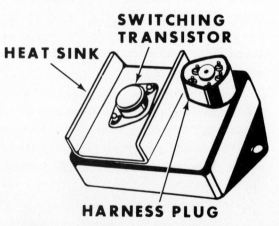

SWITCHING TRANSISTOR

HEAT SINK

HARNESS PLUG

ELECTRONIC CONTROL UNIT

ELECTRONIC IGNITION SYSTEM

Chrysler's Electronic Ignition System is designed for use on all their 1972 V-8 engines. The system is breakerless, thus eliminating the need for both breaker points and condenser.

Two short primary leads running to the magnetic distributor from a quick-disconnect, provide an easy method of identifying the ignition system. A dual-type ballast resistor is mounted on the firewall and an electronic control unit containing a switching transistor is positioned on either the firewall or fender shield. A standard ignition coil along with the necessary wiring harness complete the components of the system.

The distributor plate contains the pole piece, a permanent magnet, and a pickup coil instead of the conventional breaker points and condenser. The distributor cam is replaced with an 8-tooth reluctor.

The basic electrical law on which the electronic circuit functions is that anytime a magnetic field is broken, a proportionate voltage is induced. Each time one of the teeth of the distributor shaft-driven reluctor passes the permanenet magnet, the magnetic field is interrupted and a voltage signal or "timing pulse" is induced in the pickup coil and sent to the control unit. The switching transistor then interrupts the primary circuit thereby generating high voltage in the coil secondary windings to fire the spark plugs at the precise instant required.

The switching transistor interrupts the primary circuit for a specific length of time dictated by the electronic circuitry in the control unit. This primary circuit interruption constitutes the dwell angle. The dwell angle can be read with a dwell meter but the angle is not adjustable. No provision for dwell angle adjustment has been provided since the design of the circuitry precludes the possibility of any dwell angle change. Variations in ignition timing, subsequent to changes in dwell angle setting, have also been eliminated. Chrysler states that periodic checks of dwell angle and ignition timing settings are not necessary on the Electronic Ignition System.

Chrysler's development of the Electronic Ignition System extends the trend to breakerless ignition in an effort to assure conformance to Federal standards requiring effective emission control for a 50,000 mile interval. Since neglected breaker points are one of the most common causes of ignition misfire, which is responsible for increased exhaust emissions, their elimination relieves the possibility of the formation of objectionable emissions from this source.

Maintenance of the Electronic Ignition System is reduced to occasional inspection or testing of the spark plug cables, and replacement of the spark plugs, as required. In the event the magnet, pole piece, or pick-up coil become defective, their replacement can be performed as easily as replacing the conventional breaker points. Other system units can be similarly tested and replaced with ease.

UNITIZED OR HIGH ENERGY IGNITION SYSTEM

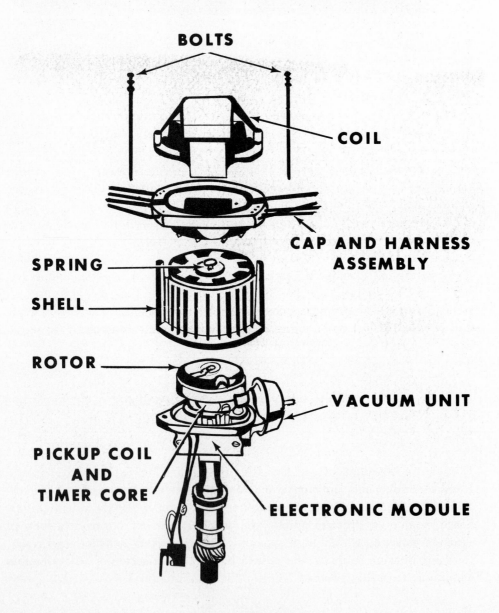

BOLTS

COIL

CAP AND HARNESS ASSEMBLY

SPRING

SHELL

ROTOR

VACUUM UNIT

PICKUP COIL AND TIMER CORE

ELECTRONIC MODULE

UNITIZED IGNITION SYSTEM
OR HIGH ENERGY IGNITION (HEI) SYSTEM

Delco-Remy introduced its Unitized Ignition System for use on the 1972 models of General Motors-built cars. The system is called unitized because the entire V-8 ignition system is built into one basic assembly—the distributor. The assembly contains the ignition coil, secondary wiring harness and cap, shell, rotor and distributor. The assembly is held together by two through bolts. This single assembly with only 9 connections takes the place of 12 components and 21 connections used in the conventional ignition system. Reducing the number of connections reduces the number of potential loose, corroded, or oxidized connections that lead to losses in ignition system performance and increased exhaust emissions.

The distributor operates on an electronically-amplified pulse, utilizing the latest high power solid-state and integrated circuitry. The conventional breaker points and condenser have been eliminated. The separate electronic amplifier unit formerly used on solid-state ignition systems has also been eliminated and has been replaced with a miniature electronic module positioned in the distributor housing.

The magnetic pickup assembly is positioned in the distributor housing just above the shaft main bearing. The assembly is designed to rotate under the influence of the vacuum advance unit. Distributor vacuum spark advance is applied in the usual manner. The timer core, mounted on the distributor shaft, is under the influence of the centrifugal weights which provide the mechanical spark advance in the conventional manner.

The magnetic pickup assembly in the distributor contains a permanent magnet, a pole piece containing internal teeth, and a pickup coil. A timer core containing projections called teeth, replaces the conventional cam and rotates inside the pole piece. When the teeth of the rotating core align with the teeth of the stationary pole piece, an induced voltage in the pickup coil sends a signal to the electronic module to interrupt the primary circuit. The high voltage induced in the coil secondary windings by this action, is directed by the rotor in the proper cable and spark plug.

The Unitized Ignition System has been designed to provide performance that conforms to the Federal emission standards requiring that emission systems effectively maintain emission control for at least a 50,000 mile interval.

TRANSISTOR IGNITION SYSTEM TESTING
BREAKER POINT TYPE

1/4" GAP   MAXIMUM

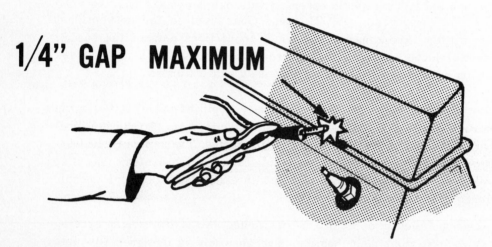

Caution: Do not draw spark near carburetor or raw gas.

HIGH RANGE OHMMETER

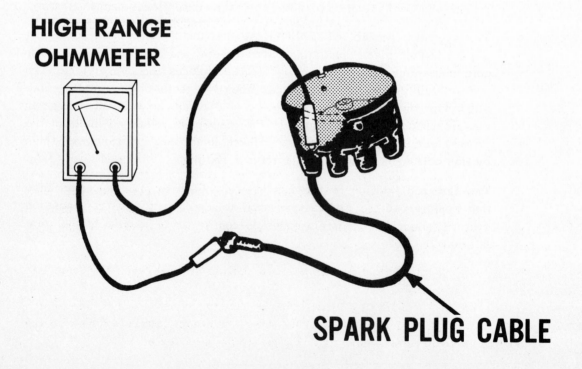

SPARK PLUG CABLE

TRANSISTOR IGNITION SYSTEM TESTING

BREAKER POINT TYPE

Due to the multitude of transistor ignition systems presently marketed, it is impossible to list a series of trouble-shooting tests that will apply equally to all systems. It is advisable, therefore, that the diagnostician furnish himself with specific test instructions on the systems he is servicing.

Before making tests on the ignition system, be sure the fuel system is functioning properly. Make checks of the carburetor choke, accelerating pump, air cleaner, fuel filter and fuel pump, as required.

The transistor ignition system may be checked using the following test sequence:

1. Remove the coil high voltage lead from the distributor cap and hold its terminal end about ¼ inch from a good ground. Crank the engine and observe the spark.

   Caution: Permit spark to jump to ground. DO NOT operate on open circuit.

   a. If a "hot" spark is present, it is an indication that the primary circuit, the coil, and the amplifier are functioning normally. Any ignition problems as misfiring, hard starting or no-start, are likely caused by conditions in the distributor cap, rotor, spark plug cables, spark plugs or ignition timing.

   b. If the spark is very weak or there is no spark, the trouble is likely in the primary circuit, the coil, or in the coil to distributor high tension lead.

2. If the condition in 1 a. is indicated, (hot spark with ignition system malfunction), proceed as follows:

   • Check ignition secondary cables. With spark plug cables removed from spark plugs and cap lifted from distributor, touch ohmmeter prods to spark plug connector and to terminal inside distributor cap for that cable. Check all cables in turn for resistance in excess of 30,000 ohms per cable. Resistance may be caused by damage in cable, in defective connection at spark plug connector, in corroded cap tower or in unseated cable in tower. Also test the coil high tension lead.

   • Check outside and inside of distributor cap for carbon tracks and cracks. Check rotor for same defects.

   Note: If either cap or rotor are defective, always replace both at the same time.

   • Check condition of spark plugs. Remove, clean, file and gap spark plugs. Replace worn or defective plugs in sets. Be sure rubber boots on cap towers and on spark plug are properly seated.

3. If the condition in 1 b. is indicated (weak spark or no spark) proceed as follows:

- Disconnect primary lead from distributor. Hold coil secondary lead about ¼ inch from a good ground and with ignition switch in "On" position, intermittently ground primary lead.

- If a hot spark occurs at the coil lead the trouble is in the breaker points. Points may be contaminated by excessive (or the wrong) cam lubricant or by oil forced into the distributor housing by crankcase pressure developed by a neglected PCV System.

- If there is a weak spark or no spark the trouble is in the amplifier or the coil. Test the amplifier ground with a jumper lead. The amplifier can be tested by unplugging the harness and contacting the amplifier connector terminals with an ohmmeter. Observe the resistance reading and reverse the ohmmeter leads. If a low resistance, or no resistance, is indicated in both tests, the amplifier has a defective transistor and must be replaced. A good amplifier is indicated by a high resistance reading in one direction and a low reading in the other.

The ignition coil may be tested by replacing it with a coil known to be good. Or the coil may be tested on a coil tester following the equipment manufacturers instructions.

Note: The coil tester must be approved for transistor coil testing or the test results will be in error. A coil may also be tested with an ohmmeter. Disconnect the coil primary leads and contact the coil terminals with the ohmmeter leads. This will check the primary coil for continuity. This test will not detect an intermittent open, however. Contacting the coil output terminal and either primary terminal will check the coil secondary winding resistance and an open circuit possibility. Compare ohmmeter readings to coil specifications.

As previously mentioned, for the best test results, follow the test procedure recommended by the manufacturer of the transistorized ignition system being serviced. Further, always observe the precautions stated relative to proper test instrument hook-up to prevent transistor damage.

TRANSISTOR IGNITION SYSTEM TESTING

MAGNETIC PULSE TYPE

TRANSISTOR IGNITION SYSTEM TESTING
MAGNETIC PULSE TYPE

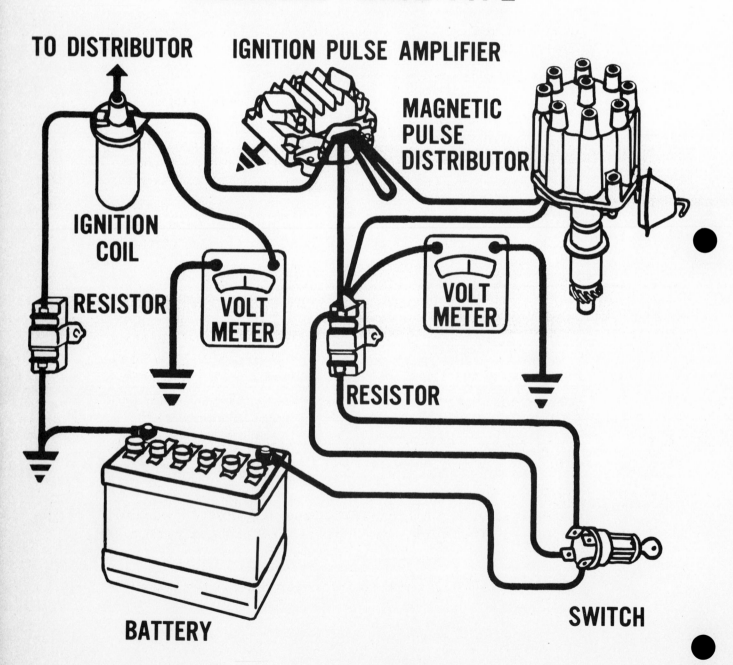

TO DISTRIBUTOR

IGNITION PULSE AMPLIFIER

MAGNETIC
PULSE
DISTRIBUTOR

IGNITION
COIL

RESISTOR

VOLT
METER

VOLT
METER

RESISTOR

BATTERY

SWITCH

TRANSISTOR IGNITION SYSTEM TESTING

MAGNETIC PULSE TYPE

Testing the magnetic pulse transistor ignition system is generally to diagnose trouble for three conditions: misfiring, surging, or no-start.

MISFIRING

Perform the following tests in the sequence listed if the complaint is misfiring.

- Check the fuel system for adequate fuel supply. Pay particular attention to fuel filters and pump pressure.

- Check the ignition timing for recommended setting.

- Remove the spark plugs. Clean, file and gap the plugs. Replace worn or defective plugs in sets.

- Check all ignition system wiring, both primary and secondary. Look for brittle or cracked insulation, loose connections or corroded terminals. Be sure high tension leads are firmly pressed into the cap towers and boots are securely seated.

- Check outside and inside surfaces of the distributor cap for cracks and carbon tracks. Also check the rotor. If either the cap or the rotor is defective, replace both.

- The pick-up coil can be tested with an ohmmeter. Disconnect the pick-up coil connector and insert the ohmmeter prods into the connector. Coil resistance should be between 300 to 400 ohms. If the meter reading is infinite, the coil is open. If the reading is low, the coil is shorted. Now touch one of the ohmmeter leads to the distributor housing. The meter reading should be infinite. If it is not the coil is grounded.

- Test the ignition coil with a tester approved for transistor coils. Follow the test procedure recommended by the equipment manufacturer.

- If all conducted tests do not locate the cause of the trouble, and the amplifier is properly grounded, replace the amplifier.

SURGE

An engine power surge condition can be prompted by reversed leads in the pick-up coil connector body. Check the lead color coding for proper connections.

Check the pick-up coil for opens, shorts and grounds as previously outlined.

Since the pick-up coil is moved by distributor vacuum advance unit action, a break in the pick-up coil wiring can cause an intermittent surging condition. To check for this condition, disconnect the distributor vacuum line and observe engine performance at idle speed. Another check for this condition can be performed by operating the distributor vacuum advance unit (on the vehicle) by applying vacuum from a distributor test bench while noting variations in the ohmmeter readings.

NO START

With a complaint of no start, hold one spark plug cable about ¼ inch from a good engine ground while cranking the engine with the ignition switch turned "On." If a spark occurs, the trouble is very likely not in the ignition system. Check out all conventional causes of engine failure to start.

If a spark does not occur, check the ignition system as follows:

● Check all the ignition system wiring as previously outlined.

● Check the distributor cap and rotor as previously outlined.

● Check system circuit continuity as follows:

a. Connect voltmeter from ignition coil positive primary terminal to ground. Turn ignition switch "On" and observe voltmeter reading. Reading should be 8 to 9 volts.

If reading is battery voltage there is an open circuit in the coil primary winding, in the resistor, or in the wiring from the coil to ground.

If the reading is Zero, there is an open in the circuit between the battery and the voltmeter connection. This part of the circuit includes the ignition switch, resistor, distributor, amplifier and connecting wiring.

b. Connect voltmeter from distributor side of resistor to ground. Turn ignition switch "On" and observe meter reading.

If reading is Zero, there is an open circuit in the resistor, in the ignition switch or in the connecting wiring.

If reading is battery voltage, there is an open circuit in the amplifier or the connecting wiring. If the wiring is not defective, replace the amplifier.

TESTING EXHAUST EMISSION CONTROL SYSTEM

ASSIST UNITS

TESTING EXHAUST EMISSION CONTROL SYSTEM ASSIST UNITS

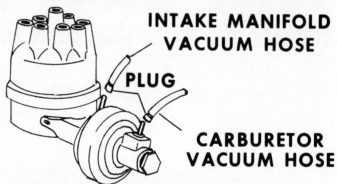

INTAKE MANIFOLD VACUUM HOSE

PLUG

CARBURETOR VACUUM HOSE

SET IGNITION TIMING
TEST CENTRIFUGAL ADVANCE MECHANISM

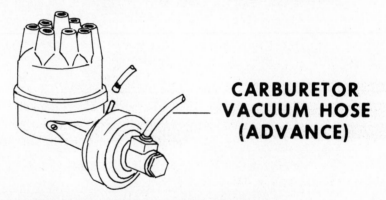

CARBURETOR VACUUM HOSE (ADVANCE)

TEST VACUUM SPARK ADVANCE

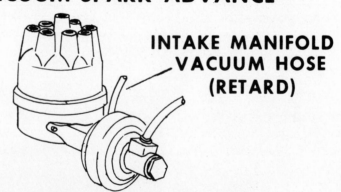

INTAKE MANIFOLD VACUUM HOSE (RETARD)

TEST VACUUM SPARK RETARD

TESTING EXHAUST EMISSION CONTROL SYSTEM ASSIST UNITS

One of the most critical factors in engine tune-up is the testing of the distributor.

If the centrifugal (mechanical) advance mechanism or the vacuum advance unit fail to function properly the ignition timing will be late, acceleration will be poor, top speed performance will be unsatisfactory and tendency toward overheating may occur. If either the mechanical or the vacuum advance systems should stick or freeze in the advance position, serious spark knock with all its engine damaging effects will result.

CENTRIFUGAL MECHANISM AND VACUUM UNIT TESTS

1. With the engine at operating temperature, connect a power timing light and tachometer to the engine.

2. Disconnect and tape (or plug) the distributor vacuum line. If the vacuum unit has a dual diaphragm or has a double-acting diaphragm, disconnect and tape both vacuum lines.

3. Idle the engine at low idle speed. It may be necessary to reduce the rpm if the idle speed is over 600 rpm as a partial mechanical advance may be introduced at this speed.

4. Check the ignition timing setting by observing the timing marks. Reset the timing to specifications as required.

5. To test the centrifugal advance mechanism accelerate the engine to 2000 rpm while observing the timing marks. Observe if a substantial amount of timing advance occured when the engine was accelerated as indicated by the movement of the timing marks. The degrees of mechanical timing advance can be accurately tested if a delay mechanism equipped-timing light is used.

6. To test the vacuum advance, lower the engine speed to 1500 rpm. While noting the position of the timing marks, connect the vacuum hose to the forward (advance) vacuum fitting. This will be the carburetor vacuum hose on dual action vacuum units. As soon as the hose is connected there should be an increase in the spark timing as indicated by the timing marks.

7. To test the vacuum retard on dual diaphragm or double-acting vacuum units, drop the engine speed to idle rpm. While observing the timing marks, connect the intake manifold vacuum hose to the rear (retard) vacuum fitting. The ignition timing should indicate a definite retard as soon as the hose is connected.

Trouble indicated in the centrifugal mechanism in Step 5 requires disassembly and servicing of the distributor advance weights and cam mechanism. Lack of timing advance in Step 6 or timing retard in Step 7 are indications of defective vacuum diaphragms. Either condition requires the replacement of the vacuum advance unit.

TESTING THE THERMOSTATIC VACUUM SWITCH
A simple procedure can be used to test the Thermostatic Vacuum Switch.

1. Connect a tachometer to the engine and bring the engine up to operating temperature. Observe the idle rpm.

2. Disconnect the intake manifold vacuum hose from the switch and plug or clamp the hose. There should be no change in the idle speed. If there is a drop in idle speed of 100 rpm or more the switch is defective and should be replaced. Reconnect vacuum hose.

3. Next, cover the radiator sufficiently to induce a high-temperature engine coolant condition as indicated by the temperature gauge or warning light. DO NOT allow engine to overheat beyond this point. If the engine idle speed has increased by 100 rpm or more by this time, the switch is performing properly. If an increase in engine idle speed has not occurred, the switch is defective and should be replaced.

TESTING THE DECELERATION VACUUM ADVANCE VALVE
The Deceleration Vacuum Advance Valve is tested by connecting a vacuum gauge to the distributor vacuum hose with a "Tee" that has the same inside diameter as the hose. Connect a tachometer to the engine and tape the carburetor dashpot plunger so that it cannot contact the throttle lever.

1. Increase engine speed to 2000 rpm and hold the speed for approximately 5 seconds.

2. Release the throttle and observe the distributor vacuum. The vacuum should increase to more than 16 inches for a minimum of 1 second and fall below 6 inches within 3 seconds after the throttle is released.

If it takes less than 1 second or more than 3 seconds to obtain the correct vacuum reading, set the valve adjusting screw. Remove the plastic cover and turn the valve adjusting screw counterclockwise (in 1/4 turn increments) to increase the time distributor vacuum remains above 6 inches. Turning the screw clockwise will decrease the time limit. A valve that cannot be adjusted to specifications should be replaced.

DISTRIBUTOR SOLENOID TESTS

DISTRIBUTOR RETARD SOLENOID

The distributor retard solenoid can be tested for proper operation as follows.

1. Disconnect and plug distributor vacuum hose.

2. Start and idle engine. Check ignition timing to specifications with power timing light. Reset as required.

3. Disconnect retard solenoid connector at carburetor.
   Note: Do NOT attempt to disconnect lead at distributor solenoid.

4. Recheck ignition timing setting. Timing should advance from initial setting and engine speed should have increased.
   Note! If timing setting has not advanced, check for good ground contact at carburetor ground switch. If ground connection is good but timing does not advance after retest, retard solenoid is defective and must be replaced.
   Note: Do not use jumper wires on solenoid or connect meters to the solenoid.

DISTRIBUTOR ADVANCE SOLENOID

Malfunction of the advance solenoid will result in hard starting. The action of the solenoid can be tested by using the following procedure.

1. Connect a tachometer to engine.

2. Disconnect and plug distributor vacuum hose.

3. Start and idle engine.

4. Disconnect solenoid lead bullet connector which is about 6 inches from distributor.
   Note: Do NOT attempt to disconnect lead at distributor housing.

5. Clip a jumper wire to the solenoid lead male connector and make-and-break contact with other end of jumper lead to battery insulated post, while observing tachometer. Engine speed should increase about 50 rpm or more each time solenoid is energized. Replace a defective solenoid.
   Note: Avoid continuous application of battery voltage for periods exceeding 30 seconds to prevent possible voltage damage to solenoid.

   Note: If Chrysler Electronic distributor advance solenoid is being tested, the single wire to distributor should be disconnected, NOT the double-bullet connector.

DECELERATION (DECEL) VALVE TESTS

MALFUNCTION SYMPTOMS

Malfunction of the deceleration valve is usually indicated by either a rough idle or a high idle speed condition.

A rough idle can be the result of a lean fuel mixture caused by a leaking valve diaphragm which permits the constant entry of additional air. If covering the small bleed hole in the bottom cover with a finger tip restores a smooth idle a defective diaphragm is indicated and the valve should be replaced.

An excessively high idle speed of approximately 1200-1300 rpm can be caused by the valve being stuck open permitting a constant draw of additional fuel from the deceleration section of the carburetor. This defect also requires valve replacement.

TEST PROCEDURE

The "decal" valve can be tested with the aid of a tachometer and a vacuum gauge.

1. Disconnect valve hose from carburetor, install a "T" fitting and connect vacuum gauge to "T". Connect tachometer to engine.

2. Start and idle engine. Observe vacuum gauge reading which should be ZERO vacuum. A vacuum reading indicates valve is stuck open.

3. Accelerate engine to 3000 rpm for a few seconds, release throttle quickly. Observe time required to drop vacuum reading to ZERO.
   Valve timing: 1600cc engine – 3 to 5 seconds
                     2000cc engine – 1½ to 5 seconds

4. If time setting is out of limits, remove cap from top cover for access to adjusting screw. Slowly turn plastic adjuster counterclockwise ½ turn to lengthen time setting, or ½ turn clockwise to shorten the time setting.

If proper time setting cannot be obtained, replace the valve.

EXHAUST GAS RECIRCULATION (EGR) SYSTEM TESTS

EXHAUST GAS RECIRCULATION (EGR) SYSTEM TESTS

Checking of the Exhaust Gas Recirculation System should be performed as part of every tune-up or at more frequent intervals if so recommended.

BUICK TEST PROCEDURE

1. Check EGR valve action by accelerating engine to 1200–1500 rpm. Valve shaft should move upward as engine is accelerated and return to downward position as engine speed returns to idle rpm.

   The action of the check valve can be observed by holding a mirror under the open lower portion of the valve body, or can be felt by inserting a finger into the body opening.

2. Apply vacuum from an outside vacuum source to the vacuum supply tube on top of the valve. Diaphragm should move to full-up position with between 8'' to 10'' of vacuum and should not leak down.

The metering valve cannot be disassembled for service. If it fails to conform to test requirements, it must be replaced.

CHRYSLER TEST PROCEDURE

The EGR System should be inspected for deposit build-up in the jet orifices at every 12,000 mile interval.

1. With the air cleaner removed and the engine NOT running, hold choke and throttle valves open.

2. With flashlight, look through carburetor throat into intake manifold floor and observe jet orifices.

   If jets are open, no service is required.

3. If jets are clogged, remove carburetor.

4. Using thin-wall socket, remove jets from manifold floor. Caution: Jets are made of anti-magnetic stainless steel and may be difficult to retrieve if dropped into intake manifold ports.

5. Clean jets in solvent until orifices are open. Do NOT use a drill or wire for cleaning as calibrated orifice may be enlarged. Increased orifice size can result in rough idle.

6. Install cleaned jets and tighten to 25 ft. lb. torque.

7. Install carburetor using new gasket, if required.

AIR INJECTION PUMP

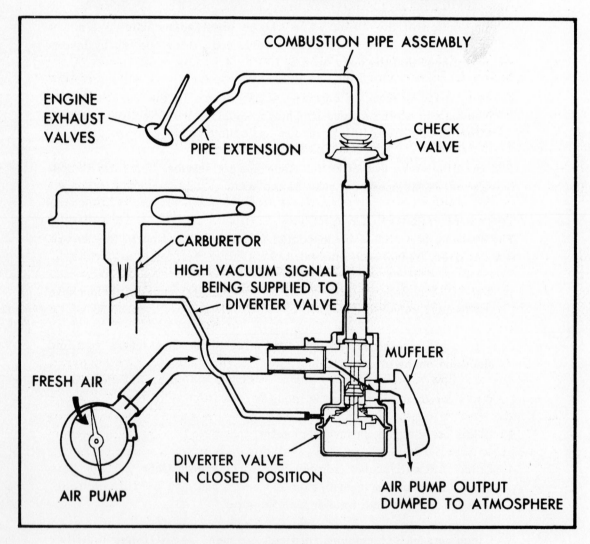

COMBUSTION PIPE ASSEMBLY

ENGINE
EXHAUST
VALVES

PIPE EXTENSION

CHECK
VALVE

CARBURETOR

HIGH VACUUM SIGNAL
BEING SUPPLIED TO
DIVERTER VALVE

MUFFLER

FRESH AIR

DIVERTER VALVE
IN CLOSED POSITION

AIR PUMP

AIR PUMP OUTPUT
DUMPED TO ATMOSPHERE

AN AIR INJECTION PUMP SYSTEM

AIR INJECTION PUMP

Another way to reduce the level of unburned hydrocarbons and carbon monoxide is to further oxidize them by injecting air into the exhaust manifold. This creates an "afterburner" condition due to the combined action of the high exhaust temperatures and the additional oxygen from the injected air. Oxidation is merely another way of saying "burning."

An engine-driven pump supplies the necessary air at a low pressure. As the air leaves the pump, it passes first through a *diverter valve* and then a *check valve* before it is injected, through small pipes, into the exhaust manifold near each exhaust valve. An 8-cylinder engine, for instance, would have eight such pipes.

The purpose of the diverter valve is to shut off momentarily the flow of air during deceleration to prevent backfiring. This function is controlled by the *change* in intake manifold vacuum. When manifold vacuum suddenly rises, as it does during deceleration, it causes a piston in the diverter valve momentarily to change position. This action "diverts" the air from the pump to the atmosphere. Proper functioning of this valve can be determined by disconnecting and then immediately reconnecting the valve's vacuum control line as the engine idles. When the hose is reconnected, the momentary action of the diverter valve should cause a puff of air to escape from the side of this valve.

The check valve or valves (two are required for V-8 engines) allow only a one-way flow of air — from the pump to the exhaust manifold. Should the pump fail, these valves keep hot exhaust gas from flowing back through the system. A quick test of these valves can be made by disconnecting the outlet hose from the diverter valve and seeing if exhaust gas flows back through the hose. Any leakage indicates a defective check valve.

THE CATALYTIC CONVERTER

The catalytic converter is basically a muffler-like device installed in the exhaust system that is designed to more fully complete the burning or oxidation of the fuel as it leaves the engine. It is these unburned exhaust emissions that are considered the major pollutants. The converter, in its present form, does not control the emissions of oxides of nitrogen. These still must be controlled by engine modification devices — principally exhaust gas recirculation and vacuum advance controls.

The converter functions as an "after-burner," promoting oxidation through the action of exhaust temperature, catalytic material and oxygen. The air supplies the necessary oxygen and is obtained from an engine-driven air pump, similar to those used with Air Injection Reaction systems.

Actually, the converter merely continues the burning process started by the engine. Even in the most efficiently burning engine, a small portion of the air-fuel charge passes through unburned, or, in other words, not fully oxidized. The unoxidized components appear as carbon monoxide and unburned hydrocarbons. The converter completes the oxidation process so that these components are "converted" to carbon dioxide and water vapor.

A catalyst is a material that accelerates a chemical reaction but without entering into the reaction or being consumed by it. Many materials can act as catalysts, depending on the particular chemical reaction involved. Most of the catalytic converters currently being produced for vehicles are of the noble metal types — that is, those using platinum and palladium. Base metal converters are also made.

Only a very small amount of catalytic material is needed in each converter. The trick is in depositing the material on a supporting *substrate* so as to expose as much of its surface as possible. A catalyst works by area, not volume. The substrate is made of aluminum oxide either in the form of "beads" or a "monolithic" surface much like a honeycomb.

To produce just one ounce of platinum and palladium requires approximately 10 to 15 tons of ore. The refining process takes approximately four months, but the yield is a metal that is 99.95% pure. The pure metal is shipped from the refiner to the suppliers of the catalytic beads or monoliths. Here, through a secret process, the highly refined metal is coated onto the substrate in a way that exposes a maximum of the metal's area. A 260 cubic inch model converter, for example, has an effective surface area equal to 59 football fields.

328

To remain effective, the converter should not be subjected to leaded fuels. This additive coats the catalytic material and renders it ineffective. However, damage does not occur immediately, but rather with prolonged or repetitive use of leaded gasoline. Occasional use of such a fuel, should a no-lead grade be unavailable, will not cause permanent damage. A slight accumulation of lead coating will "boil off" after awhile because, to a certain extent, a converter is self-purging. However, even with proper care, the catalytic material eventually will become ineffective and require replacement.

One of the claims made for the converter is that it restores lost power and economy by eliminating some of the previous emission control techniques. Since much of the job of getting rid of excessive hydrocarbons and carbon monoxide lies in the converter, engine adjustments can be put back to those that improve engine efficiency. Most authorities agree that this should mean improved economy, although they disagree as to how much. Some say the gain may be only a few per cent, others predict much higher gains.

The converter is "after the engine" and therefore has no direct influence on engine operation. This means that the driver, unless the vehicle is specifically tested for emission levels, will be unable to determine from driving whether the converter is functioning. One clue to a nonfunctioning unit is temperature. Since the converter develops heat during normal operation, an inoperative one will be considerably cooler than one that is functioning normally.

Operating temperature, in fact, has been one of the objections raised about catalytic converters. Because they are normally placed under the vehicles, it was felt they may present a hazard, particularly with regard to grass fires. However, according to converter engineers, the temperature of the outer aluminized steel shell will not exceed 250°F at the top nor go over 450°F at the bottom. This, as is stated, does not pose any fire hazard. When under-floor converters are used (a few will be located near the engine), they will generally be located under the feet of the right front passenger.

IGNITION MANUFACTURERS INSTITUTE TUNE-UP PROCEDURE

1- Cranking Voltage
2- Compression
3- Spark Plugs
4- Distributor
5- Dwell Angle
6- Ignition Timing
7- Carburetor
8- Charging Voltage

IMI TUNE-UP PROCEDURE

In our first session, we spoke about the five essential elements needed to perform quality tune-up. One of those essentials was a Specific Tune-Up Procedure.

Following a definite step-by-step procedure when tuning an engine and using the same procedure on every tune-up job is very important to consistently perform quality tune-up and effectively limit emissions.

Without a definite procedure, a tune-up man is apt to use a different approach or procedure on every engine. This leads to critical services being overlooked from time to time. It also leads to imperfect results after the tune-up is completed necessitating rework and wasted time.

Besides, if the tune-up specialist uses the same procedure repeatedly, the operation becomes "second nature" and the jobs proceed smoothly and quickly. The trouble is found the first time around which is the tune-up man's major objective. Further, the car owner can be assured his vehicle can pass a State emission test, if he is subject to a State inspection.

The Ignition Manufacturers Institute Tune-Up Procedure contains the steps which are essential to a good tune-up in the sequence in which they should be performed.

1. Cranking voltage

2. Compression

3. Spark plugs

4. Distributor

5. Dwell angle

6. Ignition timing

7. Carburetor

8. Charging voltage

SUPPLEMENTARY TUNE-UP SERVICES

1 – Inspect Drive Belts
2 – Check Manifold Heat Control Valve
3 – Service Carburetor Air Cleaner
4 – Check Fuel Filter
5 – Replace Positive Crankcase Ventilation System Valve
6 – Check Valve Clearances

SUPPLEMENTARY TUNE-UP SERVICES

Several other units on the automobile engine have a definite effect on engine operation and the limiting of emissions. The full benefits of tune-up and emission control will be realized only when these units are also functioning properly.

The following items should be checked before your tune-up operation. If their service has been neglected during normal lubrication and service intervals, their condition will undesirably influence your tune-up.

1. Inspect condition and test tension of all drive belts. Slipping belts result in: an undercharged battery even though the charging system is functioning properly; engine overheating; loss of power steering assist; loss of air conditioning efficiency; and lack of sufficient air supply from the Air Injection Reactor pump. A check of any of these systems always starts with an inspection and test of the drive belts.

2. Check manifold heat control valve action. A valve stuck in the open position will cause poor cold-engine operation and prolonged engine warm-up with extended use of the choke. A valve stuck in the closed position will result in loss of performance when normal operating temperature is reached. It can also result in burned engine valves due to an excessively lean carburetor fuel mixture.

3. Service or replace dirty air cleaners. A dirty carburetor air cleaner acts as a partial choke, upsetting the carburetor air-fuel ratio making a fine carburetor adjustment impossible.

4. Check the fuel filter. A partially clogged fuel filter will restrict the fuel supply at high speed resulting in a performance complaint. A clogged fuel filter has the same effect on engine operation as a defective fuel pump.

5. Replace the Positive Crankcase Ventilation System valve. Because most PCV-equipped engines are calibrated or adjusted to accommodate the additional air drawn from the crankcase, a clogged valve upsets the carburetor mixture causing a rough idle and making a first-class tune-up impossible.

6. Check the valve clearance adjustment when necessary. Loose tappets are noisy. Tight tappets hold valves off their seats causing engine roughness and resulting in burned out valves. Valves should be adjusted when the need is indicated.

CRANKING VOLTAGE TEST

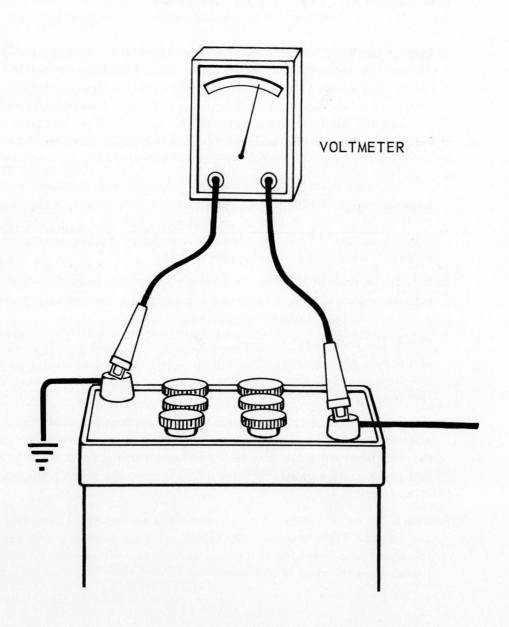

VOLTMETER

CRANKING VOLTAGE TEST

The cranking voltage test is one of the area tests used in this procedure. Area tests are used to quickly determine the condition of complete circuits. If the circuit being tested is within specified limits, it is logical to assume the individual components in the circuit must also be within limits and no further testing is necessary. If a variation from specified limits exists in a circuit test, the trouble must be located by testing each individual component in the circuit. Area testing is therefore an excellent timesaving test procedure.

A cranking voltage test quickly determines whether or not sufficient voltage is being applied to the ignition system while the engine is being cranked. Should battery voltage drop below minimum standards while the engine is being cranked, the ignition system is being starved and the engine will not start. This is a very common cause of cold weather starting difficulties.

To conduct the cranking voltage test, the voltmeter is connected across the battery. A jumper lead is used to ground the ignition primary circuit to prevent the engine from starting. With the ignition switch turned on and the engine cranking, observe the voltmeter reading and listen to the speed of the starting motor. If the voltage is less than 9.0 volts for a 12-volt battery, or less than 4.5 volts for a 6-volt battery, trouble is indicated.

The cause of the trouble may be a discharged or defective battery, defective battery cables, high resistance in battery cable connections, starting motor malfunction, a defective starting motor switch or solenoid.

COMPRESSION TEST

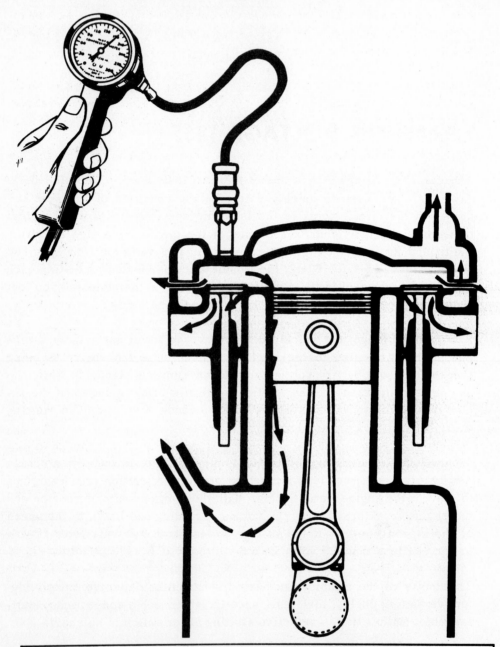

COMPRESSION TEST RESULTS						
Cylinder	1	2	3	4	5	6
Pressure (psi)	130	160	155	157	152	158

COMPRESSION TEST

The compression test reveals the mechanical condition of the engine. Any condition that permits a loss of compression pressure has a decided and undesirable influence on the engine's power output.

When conducting a compression test, you are concerned with two values: the compression pressure in each cylinder as compared to specifications, and the permissible variations of pressure existing between cylinders.

Unless the individual cylinder pressures are up to specified limits, the engine cannot be properly tuned to maximum efficiency and performance. If a greater than permissible variation of pressure exists between cylinders, a degree of engine roughness will exist that your tune-up cannot correct.

The compression test is conducted in the following manner with the engine at normal operating temperature:

● Carefully disconnect the spark plug cables from the plugs.

● Remove all the spark plugs.

● Block the carburetor linkage to hold the throttle and choke valves wide open.

● Connect remote control starting switch. Observe precautions on next page.

● Place the compression tester in cylinder No. 1 and crank the engine about four compression strokes and observe the first and fourth readings.

● Crank each cylinder the same number of revolutions and record the readings.

● Compare maximum readings and cylinder variations with specifications. Pay particular attention to the degree of variation between cylinders to make sure that a variation greater than permitted by specification does not exist. The figures in the Chart Table indicate an abnormally low reading in cylinder No. 1 since there is a greater than 10 percent variation between it and cylinder No. 2.

If compression builds up quickly and evenly to the specified pressure on each cylinder and does not vary more than the allowable tolerance, the readings are normal. The engine can be considered acceptable for tune-up.

Worn or sticking piston rings and worn cylinder walls will be indicated by low compression on the first stroke which tends to gradually build up on the following strokes. A further indication is an improvement of the cylinder reading when about a tablespoon of motor oil is added to the cylinder through the spark plug hole with an oil can. The oil temporarily seals the piston and cylinder wall clearance which results in a higher compression reading.

Valve and valve seat trouble is indicated by a low-compression reading on the first stroke and does not rapidly build up pressure with succeeding strokes. The addition of oil will not materially affect the readings obtained.

Leaky head gaskets on two adjacent cylinders will produce the same test results as valve trouble in both cylinders. An additional indication of this particular trouble is the appearance of water in the crankcase.

Carbon deposits result in compression pressures being considerably higher than specified. It is possible that carbon can hide a defect in the cylinder, as the deposit will raise the compression ratio of a cylinder to the extent which might compensate for leakage.

UNDERHOOD REMOTE CONTROL STARTING PRECAUTION

Starting in 1963, the ignition system circuitry of General Motors cars has been revised to by-pass the ignition resistor, during the cranking operation, through a contact in the starter instead of through the ignition switch as was used formerly. The change was made to overcome the possibility of a sticking switch condition which would cause distributor point failure.

The following precaution MUST be observed when connecting a remote control starting switch to a car employing the redesigned switch. Disconnect the distributor primary lead from the ignition coil and turn the ignition switch to the "ON" position while remotely cranking the engine.

Failure to observe this precaution will result in burning out the ground circuit in the ignition switch.

338

SPARK PLUGS

SPARK PLUGS

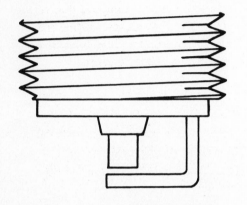

NEW PLUG ELECTRODES

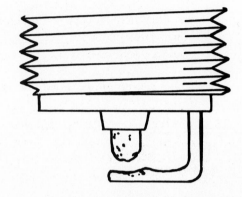

NORMAL ELECTRODE WEAR

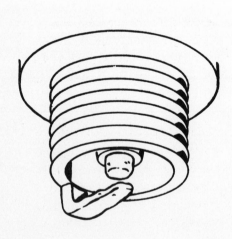

WORN OUT ELECTRODES

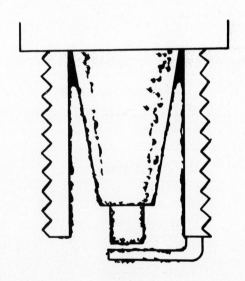

LEAD OR CARBON FOULING

SPARK PLUGS

At the completion of the compression test, the spark plugs should be serviced next.

A spark plug must operate in combustion temperatures as high as 4000 degrees F., at 1000 to 2000 sparks per minute, and withstand pressures as high as 800 pounds per square inch. After several thousand miles of service, the spark plug gap widens and the electrodes become rounded due to the combined action of intense heat, pressure, corrosive gases within the combustion chamber, and spark erosion. The plug insulator also becomes covered with carbon and lead deposits to a greater or lesser degree.

The correct spacing of the spark plug gap is very important since it influences the entire range of engine performance: starting, idling, accelerating and cruising. Further, the gap is also instrumental in controlling exhaust emissions. Uniformity of all spark plug gaps is extremely important for smooth engine operation. Always check the gap of new plugs against engine specifications before installing them.

To maintain efficient spark plug performance, plugs may be washed, cleaned, filed, inspected and regapped at regular intervals. The electrodes should be filed flat because current is emitted from a sharp corner much more easily than from a blunt surface. If this maintenance is neglected, up to 30% more voltage is required to fire the spark plugs.

Due to the extreme importance of totally efficient ignition system performance in the effective limiting of exhaust emissions many car manufacturers are recommending the replacement of spark plugs every 6,000 miles if leaded gasoline is used, and every 10,000 miles if unleaded (lead-free) gasoline is used.

If any question exists in your mind about the condition of the spark plugs you take from an engine, suggest they be replaced with new plugs. The quick starting and added acceleration the car owner receives from new plug performance and the savings he will realize in greater fuel economy, will shortly repay the cost of the new plugs. Needless to say, he will also be aiding in the fight for ''cleaner air'' since exhaust emissions are tremendously accelerated by misfiring spark plugs.

Use your specifications, and cross-reference table if necessary, for proper spark plug selection. Torque the plugs to specifications to insure proper heat transfer.

DISTRIBUTOR

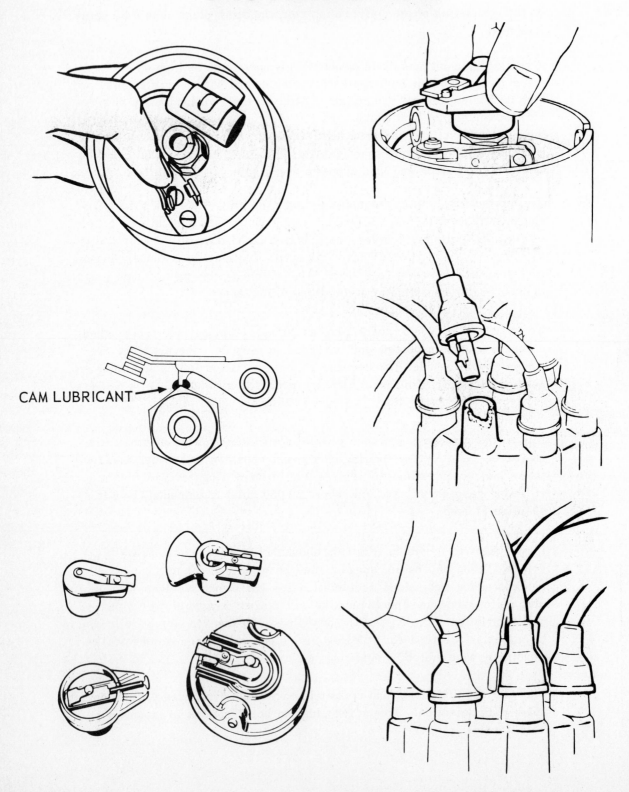

CAM LUBRICANT →

DISTRIBUTOR

The next step in the tune-up procedure is the inspection and service of the distributor components.

- Lift the cap from the distributor, gently separate the breaker points with your finger tip and inspect their condition. Points in good condition have a reasonably smooth surface with a dull grey color. If the point contact surfaces are pitted or if they have a black-and-blue appearance, they should be replaced. When replacing points always replace the condenser, too.

- Check the action of the mechanical advance weights by twisting the rotor in its direction of rotation. When released, the rotor should snap back into its released position. Failure of the rotor to snap back indicates defective springs or sticking weights.

- Wipe the distributor cam clean and apply a thin film of the proper cam lubricant as illustrated.

- Wipe the distributor cap both inside and outside with a clean cloth. Gently lift the spark plug cables from the cap, one at a time, and inspect both the condition of the tower for physical damage and the inside of the tower for evidence of corrosion. Inspect the condition of the tower terminals on the inside of the cap for erosion. Inspect the body of the cap for cracks, chips or carbon tracks. Corrosion inside the cap towers can be removed with a small round wire brush. Evidence of other physical defects necessitates cap replacement. Always replace the rotor when replacing the cap.

- Wipe the rotor clean and inspect it for an eroded terminal, weak or damaged spring and for a cracked body. If there is any evidence of erosion or damage, replace both the rotor and the cap.

- Inspect the condition of the coil. Gently pull the high tension lead from the coil tower. Inspect the tower and cap for damage and the inside of the tower for corrosion. Wipe the coil with a clean cloth. Check the condition and connections of the primary leads.

- Inspect the high tension cables. If there is any doubt of their condition, check them for continuity and resistance value with an ohmmeter. When repositioning cables in coil and distributor cap towers, make certain each cable is firmly seated.

- Inspect the condition of the rubber boots over the coil and cap towers and over the spark plugs. Boots that are cracked or carbon tracked must be replaced. If the boots are in an acceptable condition, make sure they are completely and firmly seated over the towers and spark plugs.

A few rules to follow during your distributor service that will assist you in performing a quality tune-up are:

- When installing new breaker points pay particular attention to spring tension and point alignment. Then clean the points with solvent and lintless tape.

- When the breaker points are replaced, wipe the breaker cam clean and apply a thin film of specially compounded cam lubricant to the cam.

- Never file pitted breaker points, except as a road service expedient. Point pitting is a form of electric weld and is very hard. Points that need filing, need replacing.

- Always replace the condenser when replacing the breaker points.

- Always check and reset the voltage regulator setting after replacing burned points to prevent a reoccurrence of the trouble which was caused by high voltage.

- Whenever either a distributor cap or a rotor is found defective, replace both.

- Always handle high tension cables gently to prevent internal damage which will introduce high resistance or lack of continuity.

BREAKER POINT WEAR PATTERNS

BREAKER POINT WEAR PATTERNS

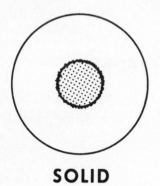

SOLID

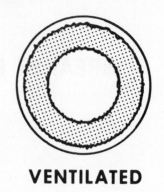

VENTILATED

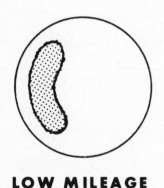

LOW MILEAGE

HIGH MILEAGE

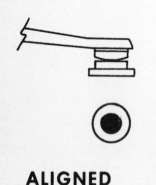

ALIGNED

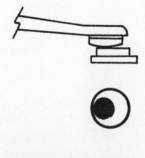

MISALIGNED

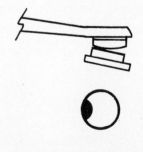

MISALIGNED

BREAKER POINT ALIGNMENT

Chart No. 122

BREAKER POINT WEAR PATTERNS

Breaker point inspection and/or replacement is a critical part of every engine tune-up. Because of the importance of this operation, it is essential that breaker point wear patterns be understood so that the true point condition can be properly appraised.

There are essentially two types of breaker points:

1. Pivoted solid contact type.

2. Pivoted ventilated type. These may be circle vented or crosscut vented.

Breaker point sets are made in both two-piece units and one-piece assemblies. If the point set has two pieces the insulated point is designed with a bushing to be positioned on the point pivot post attached to the breaker plate. The one-piece assembly is mounted on a sub-plate which basically contains its own pivot point. The pivotless breaker set is riveted into a one piece assembly and the action of the distributor cam causes the insulated breaker arm to bend or flex rather than pivot about a post as on pivoted point sets.

The breaker point design feature as, pivoted or pivotless, solid contact or ventilated type, are the factors that determine the breaker point wear pattern. The normal wear pattern of each type is illustrated on the chart. Pivoted type point patterns, besides revealing the condition of the points also indicates whether the points have been properly aligned. An important factor is that the pivotless point set has a normal low-mileage pattern that might be misinterpreted as a condition of point misalignment if this condition is not understood.

Regardless of the type of breaker point design, the appearance of the contact surfaces is much the same for all. Points that have been in operation for several thousand miles of vehicle operation have a frosted or roughened appearance.

This does not mean the points are worn out. The roughened surfaces between the contacts match or mate each other so that a large contact area is maintained and the points will continue to provide satisfactory service if not disturbed.

If an insulating oxide or dirt collection on the points interferes with proper electrical point contact, a few strokes with a clean, fine-cut ignition file will possibly dress them for reasonable contact. It should be remembered, however, that the pit build-up on the point contact surfaces is actually a very hard electric weld. No amount of filing can dress these points to a like-new surface. If the pitting appears excessive, only point set replacement will restore efficient ignition performance.

There are few cautions relative to servicing **used** breaker points.

NEVER attempt to clean pitted points by drawing emery cloth, sandpaper or tape through the points. Bits of grit, lint or fibre will stick to the pits and insulate the points from each other causing total ignition failure.

NEVER realign used points because the wear pattern indicates the points have been operating in a misaligned condition. Upsetting the mated surfaces that formed between the points will result in practically a total loss of point contact. Realigning used points cause small point **contact** areas to carry all the primary current load. The points will quickly overheat and total point failure will shortly result.

NEVER use a feeler gauge to regap used points. The use of a dwell meter is the only recommended method of accurately setting pitted points. In fact, setting breaker points by adjusting the dwell angle is the superior method to adjusting gap with a feeler gauge on both new and used breaker points.

BREAKER POINT COLOR

The color of used breaker points is an excellent indication of the condition under which they have been operating.

- Blue color. Indicates electrical trouble such as high voltage, excessive primary current flow or defective condenser.
- Black color. Indicates mechanical trouble caused by dirt, oil or grease on the contacts. The wrong cam lubricant or a plugged PCV valve inducing high crankcase pressure can be the cause.
- Grey color. Grey is a normal operating color and is usually accompanied by a slightly frosted and roughened texture which is an acceptable condition.

Before performing the distributor service part of your tune-up, be sure to correct the electrical or mechanical trouble indicated by the color and condition of the used points. This will assure that your tune-up will provide thousands of miles of trouble-free engine performance.

DWELL ANGLE

DWELL ANGLE

DWELL VARIATION

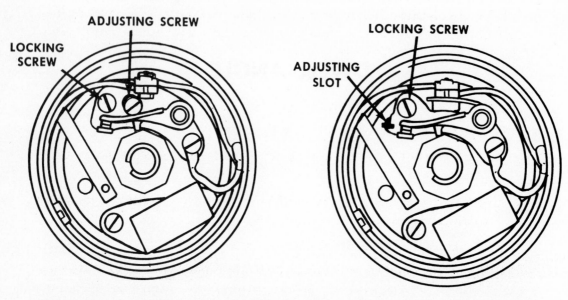

LOCKING SCREW

ADJUSTING SCREW

LOCKING SCREW

ADJUSTING SLOT

Breaker Point Locking and Adjusting Methods

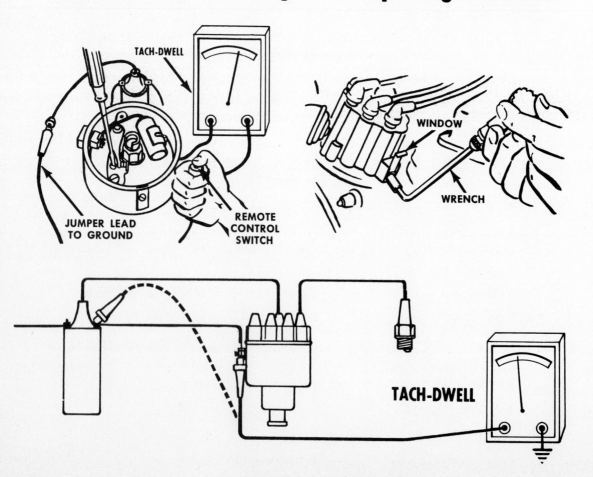

TACH-DWELL

WINDOW

JUMPER LEAD TO GROUND

REMOTE CONTROL SWITCH

WRENCH

TACH-DWELL

DWELL ANGLE   DWELL VARIATION

SINGLE BREAKER POINT ADJUSTMENT

After the distributor components have been checked and/or replaced, the breaker point dwell angle should be set to specifications. A dwell meter must be used for this operation.

To set the dwell angle on distributors having a solid cap, the following method applies:

- Connect the dwell meter.

- Connect the remote control switch observing precaution under Compression Test.

- Pull the coil lead from the distributor cap center tower and ground it with a jumper lead.

- Lift the cap from the distributor and remove the rotor.

- Disconnect and plug the vacuum hose(s) on Ford, Autolite, Prestolite and Chrysler-built distributors. Also electrically disconnect the distributor solenoid at the carburetor connector on Chrysler-built engines that are solenoid equipped.

- Turn on the ignition switch.

- Loosen the point lock screw slightly, crank the engine while slowly turning the point adjusting screw, or shifting the grounded point plate, until the specified dwell angle is indicated on the dwell meter. Then tighten the point lock screw. Recheck setting and turn off ignition.

- Remove the jumper lead, press the coil lead firmly into the distributor cap tower and seat the rubber boot.

- If the points are new, clean them with solvent and lintless tape. If the points have been used, **do not** attempt to clean them.

- Wipe distributor cam clean and apply light film of cam lubricant.

- Place the rotor and the cap on the distributor. Press each cable down firmly into the coil and cap towers and make sure each rubber boot is properly seated.

- Reconnect distributor vacuum hose(s) and distributor solenoid lead.

- Check the ignition timing and reset it as required.

If the engine is equipped with an external adjustment-type distributor (with a window), setting the dwell angle is a simple operation. It is performed in the following manner with the engine running:

- Connect the dwell meter.

- Start and idle the engine.

- Lift the window on the side of the distributor cap.

- Insert "hex" wrench or adjusting tool into the head of the point adjusting screw.

- Slowly turn the wrench while watching the dwell meter until the specified dwell angle is indicated on the dwell meter.

- Withdraw the wrench and close window **completely**. The adjustment is self-locking.

- Check the ignition timing and reset it as required.

The use of a dwell meter for setting dwell angle rather than using a feeler gauge for setting point gap has other advantages in addition to its accuracy as a method of point adjustment.

1. A distributor resistance test, another of the area tests, can be conducted by merely hooking up the dwell meter and turning on the ignition switch. With the points closed the existance of excessive resistance anywhere in the ignition primary circuit will be indicated on the meter. By working through the primary circuit with one lead of the meter, the trouble can be quickly isolated.

2. A dwell variation test can be easily and quickly made after the dwell angle has been set. If the dwell setting varies more than 3 degrees, or as specified, as the engine is slowly accelerated from idle to 1500 rpm, it indicates that either the distributor shaft bushings are worn, the distributor cam lobes are worn, that variations exist in cam lobe accuracy, that the distributor shaft is bent or there is primary circuit resistance quite likely in the distributor pigtails or in a loose or wobbly breaker plate.

3. The dwell meter also provides the only accurate method of setting breaker points that have been in use. The pitting that normally occurs on breaker points eliminates the use of a feeler gauge as an accurate method of resetting point spacing. **Always** use a dwell meter rather than a feeler gauge when readjusting used breaker points.

Note: The breaker point dwell angle must **always** be set before the ignition is timed. If the engine is timed first and the dwell setting is subsequently changed, the ignition timing will deviate from the specified setting.

The importance of applying a thin film of the recommended cam lubricant to the cam at every breaker point service can again be emphasized. A dry, rusty or pitted cam will quickly wear away the breaker point block thereby changing the dwell setting. The ignition timing setting changes in direct proportion to the dwell change thereby prompting late ignition, hard starting, poor low-speed performance and increased fuel consumption. After setting the dwell always wipe the cam clean and apply a thin film of cam lubricant. This habit will keep your tune-ups perfectly tuned for thousands of miles. If the distributor is equipped with a cam lubricator, rotate the wick or change the lubricator position (end-for-end) to present a new wick contact surface. If wick is dry, replace it, NEVER lubricate it.

DUAL BREAKER POINT ADJUSTMENT

Dual breaker points are set to specifications using a dwell meter and following this procedure:

- Connect the dwell meter.

- Connect remote control starting switch observing percautions.

- Pull coil lead from distributor cap center tower and ground it with a jumper lead.

- Lift cap from distributor and remove rotor.

- Block open one set of contacts with a piece of insulating material.

Note: The insulating material should be clean and hard preferably a piece of fibre insulation. DO NOT use a fuzzy, soft material as a paper match holder.

- Turn on the ignition switch.

- Loosen point lock screw, on opposite point set, just enough so that the stationary plate is moved with a slight drag. This will assure more precise adjustment.

- Crank engine and set breaker points to specified dwell angle for single point set. Tighten lock screw and recheck dwell setting.

- Remove insulating material from blocked points and insert insulator between adjusted points.

- Lightly loosen lock screw of second point set.

- Crank engine and set breaker points to specified dwell angle for single point set. Tighten lock screw and recheck dwell setting.

- Remove insulating material from points. Crank engine while observing total dwell reading. Total dwell must be within limits specified. If total dwell is out of limits, readjust individual settings.

- Wipe cam clean and apply a thin film of cam lubricant.

- If breaker points are new, clean them with a cleaner-type solvent and lintless tape. If the points are used, do NOT attempt to clean them.

- Replace rotor and cap on distributor. Reseat coil lead in distributor cap. Make sure each cable is firmly seated in distributor cap towers and each boot is properly seated.

- Check ignition timing and reset as required.

DUAL BREAKER POINT DWELL ANGLE SETTINGS

The specification for the dwell angle setting of individual point sets in dual assemblies may be a specified angle as: each set 26° to 28° with total dwell angle of 33° to 34°. Or the setting may be specified as: total dwell angle 30° to 33°, each set with equal dwell.

In either event, follow the specifications precisely for an accurate dwell setting.

IGNITION TIMING

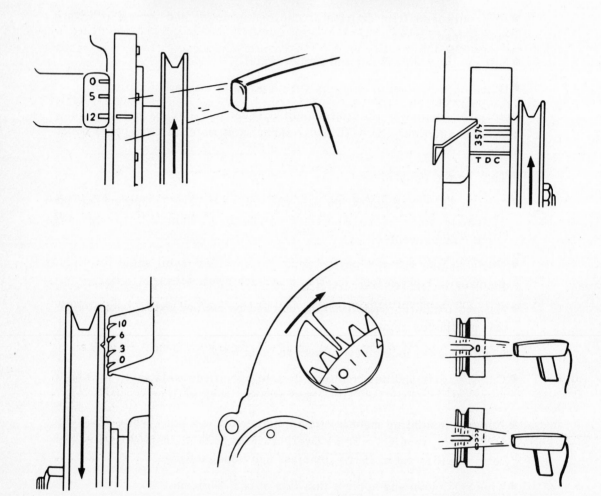

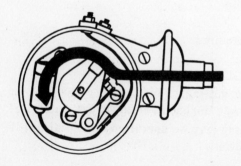

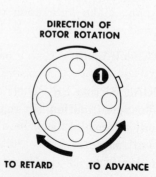

DIRECTION OF
ROTOR ROTATION

TO RETARD   TO ADVANCE

IGNITION TIMING

Correct ignition timing is one of the most important factors relative to efficient and economical engine operation. If the initial timing setting is not correct, the entire range of the spark advance curve will be out of limits.

Ignition timing is checked and set with the aid of a power timing light. The light is energized by battery current and triggered by the voltage of the spark plug to which it is connected, usually the plug in cylinder No. 1.

The procedure for ignition timing an engine is as follows:

- Locate the timing marks on the crankshaft pulley, harmonic damper or flywheel. If they are not readily visible, wipe them with a cloth and mark them with chalk or paint.

- Operate the engine until normal temperature is reached.

- Stop the engine and connect a tachometer and a timing light. Disconnect and tape the distributor vacuum hose. If there are two hoses, disconnect and tape both hoses.

- Start and idle the engine. The light will flash each time the No. 1 spark plug fires.

- Operate the engine at the specified speed and aim the timing light at the timing marks.

  Caution: Be very careful of the revolving fan blades.

- Reset ignition timing if timing mark appears on either side of reference pointer. Reconnect distributor vacuum hose(s).

The ignition timing is adjusted by slightly loosening the distributor hold-down screw and slowly turning the distributor body against rotor rotation to advance the timing or with rotor rotation to retard the timing. When the specified mark is aligned with the pointer, securely tighten the distributor hold-down screw. Then recheck the alignment of the timing marks.

The chart illustrates a method of easily determining the direction of rotor rotation without removing the distributor cap or cranking the engine by merely observing the position of the vacuum advance unit on the distributor body. Rotor rotation can be determined by visualizing an arrow passing through the vacuum line and around inside the distributor cap. As illustrated, the rotor turns counterclockwise. Had the vacuum advance unit been positioned below the centerline of the distributor, rotor rotation would have been clockwise.

The timing mark should appear steady as the light flashes. If the mark appears to "fan out" or "wander" as the light flashes, trouble is indicated. This condition can be caused by pitted breaker points, misaligned points, improper point spring tension, loose or worn breaker plate, worn distributor shaft, worn distributor shaft bushings, or excessive lash anywhere in the distributor drive mechanism. These conditions must be corrected before an engine can be properly tuned.

After the initial timing is set, slowly accelerate the engine to approximately 1500 rpm while observing the timing mark with the light. The timing mark should move steadily away from the pointer. This is an indication that the spark advance mechanism is in operation. If there is little or no indication of spark advance, the distributor should be removed from the engine for a complete test.

CONNECTING THE TIMING LIGHT

The power timing light may be designed with a pick-up unit which is placed around the spark plug cable making an easy connection. In the event a timing light is used that must be clipped to the No. 1 spark plug and the plug is not readily accessible, the No. 1 tower in the distributor cap may be used with the aid of an adapter. The spark plug in the companion cylinder to No. 1 may also be used.

The companion cylinder to No. 1 can be quickly determined by the engine's firing order:

- In a 4-cylinder engine, the companion cylinder is the third cylinder in the firing order.

- In a 6-cylinder engine, the companion cylinder is the fourth cylinder in the firing order.

- In an 8-cylinder engine, the companion cylinder is the fifth cylinder in the firing order.

The distributor cap tower of the companion cylinder may also be used if desired.

Any engine operating on the four-stroke cycle principle can be ignition timed from four locations.

NEVER puncture the spark plug cables to hookup a timing light or other test equipment. Use the proper adapter.

NEVER power time an engine. Always use a timing light.

IGNITION TIMING OF EXHAUST EMISSION CONTROL
SYSTEM EQUIPPED ENGINES

Setting the ignition timing on engines equipped with an Exhaust Emission Control System is performed in the same manner as setting the timing on conventional engines. Instructions for removing the distributor vacuum line and taping the manifold opening during timing setting and the caution about using a timing light adapter rather than puncturing secondary cable insulation, also apply to engines with California exhaust control systems.

The important factors relative to setting ignition timing on engines equipped with exhaust control systems are that the correct timing specification be used and that the timing is accurately set.

USE CORRECT TIMING SPECIFICATIONS

Usually the specifications for standard engines are given preference in the listing. California specifications may be covered with an asterisk or by footnotes. It is easy to misinterpret these specifications so use care when selecting the timing specification for the engine being tuned.

SET THE TIMING CORRECTLY

The ignition timing on many engines equipped with exhaust emission control systems is set at Top Dead Center or After Top Dead Center. This is contrary to conventional timing practice. The timing marks on some engines have been altered to indicate both Before Top Dead Center and After Top Dead Center positions. Some timing marks are scribed with three letters, one above the other. They are A, O, and R. The A means Advance or Before Top Dead Center position. The O stands for Zero or Top Dead Center position. The R means Retard or After Top Dead Center. Each line scribed on the marker stands for 2 degrees of adjustment.

American Motors uses a similar marking in which the letter B signifies the Before Top Dead Center portion of the marker.

Chrysler Corporation cars are clearly marked with the words "Before" and "After" with a O (Zero) or Top Dead Center position set between the Before and After markings.

Ford Motor Company cars, equipped with Thermactor Emission Control Systems, are not timed later than Top Dead Center so the conventional Ford timing marks have been retained.

After setting the basic ignition timing, test the distributor centrifugal and vacuum advance mechanisms. Be sure to test both the spark advance and the spark retard action on dual-action vacuum units.

POSITIVE CRANKCASE VENTILATION SYSTEM TESTS

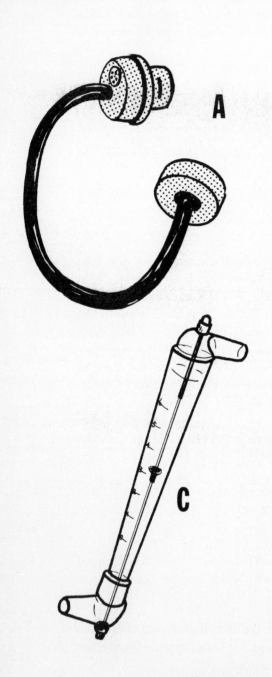

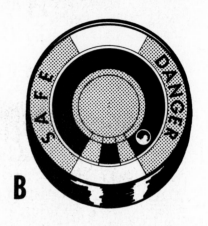

B

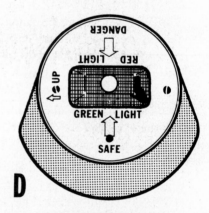

D

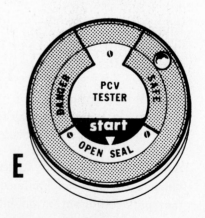

E

VARIOUS TYPE TESTERS

POSITIVE CRANKCASE VENTILATION SYSTEM TESTS

Several gauges are marketed for testing the operation of the crankcase ventilating system. The tests performed are essentially checks on the operation of the PCV valve. The testers are all basically pressure sensing devices. Their function is to check the degree of air flow and blow-by circulation through the crankcase while the engine is idling.

One tester (A) is equipped with a selection of rubber adapters to plug engine breather openings as the oil dipstick tube or the dual breather cap used on some V-8 engines. A selector knob on the bottom of the tester is adjusted to match the air flow rate of the valve on the engine being tested. With the engine idling, and at normal operating temperature, the breather cap is removed from the rocker arm cover and the tester adapter is pressed into the breather cap opening. With the tester held upright, the color indicator can be viewed through the tester window. The green color indicates proper air flow circulation so the PCV system is functioning properly. A yellow color means the system is partially plugged or the crankcase is not properly sealed. The red color indicates a blocked system which generally means a plugged valve. In the case of a worn engine, blow-by past the rings may be so excessive that it is beyond the capacity of even a properly operating PCV system to handle. In this event the engine cannot be tuned with any degree of success.

Another tester (C) differs in design from the other gauges. A hose connected to the PCV system is attached to the bottom of the tester. A hose leading to the carburetor or intake manifold is attached to the top of the tester. A free-sliding indicator is mounted on a rod housed inside the tester. The degree of vacuum passing through the system is indicated by the reading on the side of the tester body opposite the position of the sliding indicator.

The other testers are variations of the same principle. The crankcase breather is removed and the tester is firmly seated in the breather cap opening. Vacuum or pressure in the system causes the ball in the tester to indicate a safe or danger position. One of the testers (E) also indicates an "Open Seal" position which warns of possible leaks in the PCV system.

Full operating instructions are included with every tester.

When tests and service of the PCV system is not performed at every tune-up, there are several undesirable effects other than air pollution.

- Rough engine idle. A plugged or restricted PCV system upsets the carburetor air/fuel ratio. With the idle mixture unbalanced, a rough idle and a tendency to stall condition exists.

- Increased fuel consumption. Because of the unbalanced air/fuel ratio there is a proportionate increase in fuel consumption. The reduction in air supply will produce a "partial choke" condition.

- Loss of engine performance. Spark plug gas fouling associated with rich carburetor mixtures, is quickly reflected in loss of acceleration, misfiring and generally poor engine performance.

- Increased oil consumption. Crankcase pressure developed by a blocked system can force oil past front and rear main bearing seals, past rocker arm and oil pan gaskets and out the crankcase dipstick tube. A PCV valve stuck in the open position can cause oil to be pulled out of the oil pan into the engine. This condition will result in a depleted oil supply and a ruined engine.

- Motor oil sludge build-up. Unvented moisture vapors and blow-by gases condense and settle in the motor oil resulting in sludge formation robbing the engine of its vital lubrication.

- Premature engine wear. Motor oil contamination destroys the additives in the motor oil causing acid etching and rusting of critical high-polished internal engine surfaces. This condition results in accelerated engine wear. Cam lobes and hydraulic valve lifters are particularly effected by this condition.

- Crankcase odors. Fumes trapped in the crankcase give off a strong odor of hot oil that often finds its way into the car. These odors are sickening and nauseating to most people. Crankcase odors in the car is one of the first indications of a neglected PCV system.

No professional tune-up can be performed without including the testing and servicing of the PCV system and replacing the valve as required.

EXHAUST EMISSION CONTROL SYSTEM TESTS

CLEANER AIR SYSTEM TYPE (CAS)

EXHAUST EMISSION CONTROL SYSTEM TESTS

CLEANER AIR SYSTEM (CAS)

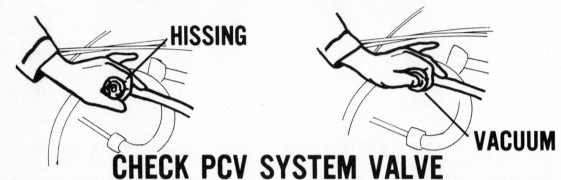

HISSING

VACUUM

CHECK PCV SYSTEM VALVE

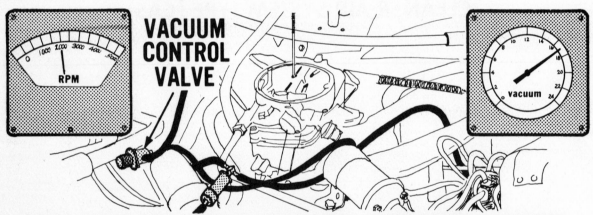

VACUUM CONTROL VALVE

RPM

vacuum

TEST DISTRIBUTOR VACUUM CONTROL VALVE

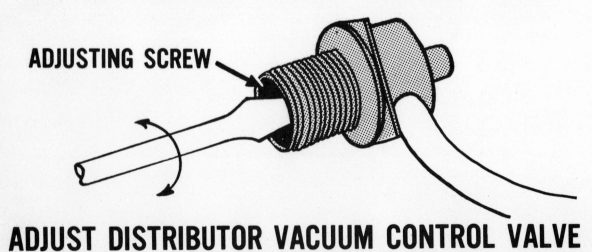

ADJUSTING SCREW

ADJUST DISTRIBUTOR VACUUM CONTROL VALVE

EXHAUST EMISSION CONTROL SYSTEM TESTS

CLEANER AIR SYSTEM (CAS)

Chrysler Corporation recommends the Cleaner Air System Exhaust Emission Control System be tested every twelve months to make certain that the system is maintaining the exhaust emissions to the specified low level.

The following tune-up procedure should be conducted in the sequence listed:

1. Spark plugs. Clean, file, gap and torque to 30 foot pounds using new gaskets. Replace worn or defective plugs, in sets.

2. Distributor. Inspect points for pitting, alignment and block wear. Replace points as required. Point replacement interval should not exceed 24,000 miles. Clean and lubricate cam and wick. Set dwell to specifications with distributor vacuum hose disconnected and plugged and distributor solenoid disconnected at carburetor, if so equipped.

3. Secondary cables. Every three years, or whenever misfiring occurs on hard acceleration, check resistance of cables with an ohmmeter. Replace cables and distributor cap if more than 30,000 ohms resistance is present, or 50,000 ohms on cables over 25 inches long. If cap is replaced, replace rotor also.

4. Coil. Inspect coil for oil leakage or carbon track flash-over.

5. Battery. Test specific gravity and add water as required. Clean battery top, cables and clamps. Coat terminals with light grease to retard corrosion collection.

6. Carburetor choke. Remove air cleaner and squirt carburetor cleaner on choke shaft ends while hand operating choke. Choke valve must work freely.

7. Air cleaner. Clean element throughly with gentle air stream. Replace element every two years or sooner, as required. If element has outer wrapper, wash wrapper in kerosene or solvent and blot or shake dry. Do not wring or twist the wrapper. Oil bath air cleaners should be inspected every six months, throughly cleaned once a year.

8. Manifold heat control valve. Check for free operation. If sticking, flush shaft ends with solvent while hand operating valve. Replace defective valves.

9. Positive Crankcase Ventilating System. Inspect PCV system every six months, replace valve every 12 months. Remove valve and hose assembly from rocker arm cover. If system is clear and valve is working, a hissing noise will be heard. A strong vacuum will be felt when finger is placed over valve inlet. Replace clogged valve, service hoses and carburetor passages as required.

10. Engine idle adjustment. With engine of operating temperature, tachometer connected, transmission in Neutral, air conditioning unit turned OFF, set idle rpm to specifications.

11. Ignition timing. Disconnect distributor vacuum line and tape manifold opening. Set timing to CAS specifications. Reconnect vacuum line. Timing should remain where set with line disconnected.

12. Carburetor air/fuel ratio. Combustion analyzer reading for CAS equipped engines is 14.2 to 1 ratio. Adjust carburetor as required by turning each mixture screw just 1/16 turn at a time. Wait 10 seconds before reading meter. Readjust idle rpm as required. Readjust mixture screws individually if idle is rough but air/fuel ratio must remain between 14.0 to 14.2.

13. Distributor vacuum. Connect vacuum gauge into distributor line with "T" fitting. With a clamp, shut off vacuum hose between control valve and intake manifold. Remove vacuum hose from distributor vacuum advance unit and clamp end of hose shut. With engine idling, vacuum reading should be between Zero to 6 inches. If reading is higher than 6 inches, recheck timing, air/fuel ratio and idle speed. Remove clamps and reconnect hoses.

14. Vacuum control valve. Increase engine speed to 2000 rpm and hold speed for about 5 seconds. Release throttle and observe vacuum gauge. Distributor vacuum should increase to about 16 inches for a minimum of one second. Vacuum should then fall below 6 inches within 3 seconds after throttle release.

    If time is less than one second, remove valve cover and turn control valve adjusting screw counter-clockwise to increase time limit. If time required was more than 3 seconds, turn adjusting screw clockwise to decrease time limit. One complete turn of the adjusting screw will change the valve setting approximately 1/2 inch of vacuum. A valve that cannot be adjusted must be replaced.

15. Inspect Exhaust Gas Recirculation System and service as required.

16. Road test. Road test for proper performance in all driving ranges. Correct deficiencies but all settings MUST be within CAS specifications.

EXHAUST EMISSION CONTROL SYSTEM TESTS
AIR INJECTION TYPE

EXHAUST EMISSION CONTROL SYSTEM TESTS

AIR INJECTION TYPE

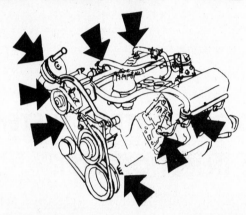

TYPICAL CHECK POINTS

- CHECK DRIVE BELT CONDITION AND TENSION
- CHECK PUMP AIR FILTER
- CHECK CONDITION AND FIT OF HOSES
- CHECK CONDITION OF FITTINGS TO AIR MANIFOLD
- CHECK CONDITION OF FITTINGS TO EXHAUST MANIFOLD
- CHECK OPERATION OF DIVERTER AND CHECK VALVES
- CHECK PCV VALVE
- CHECK IGNITION TIMING
- SET ENGINE IDLE SPEED

EXHAUST EMISSION CONTROL SYSTEM TESTS

AIR INJECTION TYPE

There is a definite relationship between engine tune-up and objectionable exhaust emissions. If the air injection system seems to be malfunctioning, it is advisable to check out the steps of a tune-up that effect the carburetor air/fuel ratio and the complete burning of the fuel charge before testing or servicing the air injection system.

If the engine is properly tuned and the exhaust emissions are in excess of the permissible amount, inspect the emission control system as follows:

● Check drive belt condition and tension. Inspect the condition of the belt for wear, cracks, oil soaking, stretching or glazed condition. Test belt tension and set to specified tension. Check for misaligned pulleys.

● Check pump air filter. Air filter element must be replaced at recommended interval, generally 12,000 miles; sooner if vehicle is operated under dusty conditions. A centrifugal filter fan is used on late-model pumps.

● Check air pump by removing air outlet hose. Accelerate engine to approximately 1500 rpm and check air flow from the hose. If air flow increases, pump is functioning properly. If air flow is lacking, or does not increase when engine is accelerated, the pump or pressure valve is defective.

● Check the diverter valve by disconnecting signal line at valve. With engine running, a vacuum signal must be present at hose. Reconnect hose.

● Hold hand against exhaust ports on muffler. With engine idling, no air passage should be felt. Accelerate engine momentarily. A blast of air should be vented through the exhaust ports.

● Inspect check valve by disconnecting air hose(s) from check valve(s). Inspect position of plate inside valve. It should be positioned against valve seat, away from the air manifold. Press down on valve plate, it should return freely to its seat when released.

● With hose(s) still disconnected, start and idle engine. Accelerate engine to about 1500 rpm and feel for gas leakage at check valves. Replace defective valves that leak. Note: Flutter of valve plate at idle speed is normal due to exhaust pulsations in the manifold.

● Check backfire suppressor valve on engines so equipped. This valve was used on early model air injection systems. Disconnect air pressure hose from backfire suppressor valve and plug. Connect vacuum gauge at point of air hose disconnect. Start and idle engine. Vacuum gauge reading should be ZERO. If vacuum is indicated, valve is defective and should be replaced. Open and close throttle suddenly. Vacuum gauge indicator should move upscale rapidly and slowly settle back to ZERO. Repeat test several times Replace valves that do not conform to test procedure.

- Check condition and fit of hoses. A burned or baked hose is an indication of a defective check valve. Check for air leaks in hoses and around clamps. With engine idling, move hand along each hose to feel for leaking air. A soapy water solution may also be used to detect air leaks.

- Check condition of fittings at air manifold(s). Check for air leaks at hose connection; check valve(s) and at air nozzle connections.

- Check condition of fittings at air tube injection nozzles at exhaust manifold.

- Test the Positive Crankcase Ventilating System. Replace valve and service hoses as required.

- Inspect and test Exhaust Gas Recirculation System, if engine is so equipped.

- Check ignition timing. Use a power timing light and set ignition timing to setting specified. Be sure the specifications apply to exhaust emission controlled engines if the car is a 1966 or 1967 model since engines both with and without emission controlled systems were built these years and the tune-up specifications were different in many instances.

- Set engine idle speed. With tachometer connected, set engine idle speed to recommended rpm.

Whenever the exhaust manifold(s) is removed, inspect the stainless steel air injection tubes for carbon build-up and for a warped or burned condition. Use penetrating oil when working the tubes free of the manifold.

AIR INJECTION PUMP

The air injection pump is not an entirely noiseless accessory. A noise that rises in pitch as the engine is accelerated is a normal sound. A chirping sound at idle or low speed is caused by contact of the pump vanes with the inside of the pump housing bore. New or rebuilt pumps have a tendency to chirp until the pump vanes have seated themselves. A squealing noise on engine acceleration is caused by a slipping drive belt. Adjust tension or replace the belt as required. Check that the pump rotates freely. A frozen pump will cause belt squeal since the pump pulley will not turn. Check the alignment of all drive belt pulleys. A pump noise that persists that cannot be corrected on the car may be caused by bearing, vane or rotor wear. Remove and repair the pump or replace it.

CAUTION –

The air pump housing is made of aluminum. Do NOT press excessively on the housing when setting belt tension. Hand pressure may be sufficient. Do NOT place the pump housing in a bench vise when servicing the pump.

TESTING HEATED CARBURETOR AIR SYSTEMS

HEATED CARBURETOR AIR SYSTEM TESTS

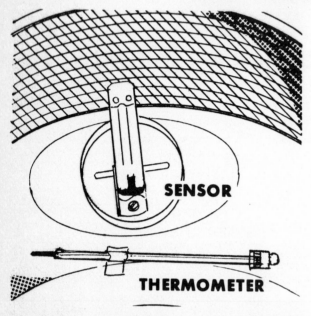

TESTING TEMPERATURE SENSOR

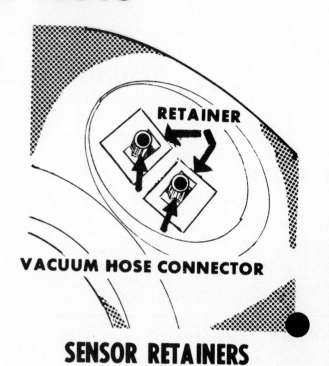

SENSOR RETAINERS

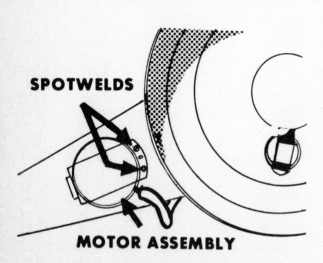

VACUUM MOTOR SPOTWELDS
GM type

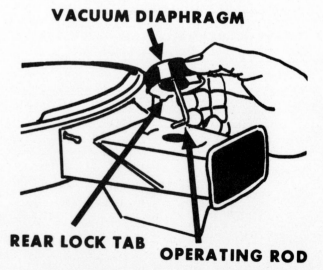

VACUUM MOTOR REMOVAL
Chrysler type

TESTING HEATED CARBURETOR AIR SYSTEMS

There are certain recognizable symptoms that arise when the carburetor heated air system malfunctions. A cold engine will idle roughly and performance will be erratic due to the fact that cold air, rather than warm air, is entering the carburetor. On the other hand, poor high-speed performance and power loss will be experienced if warm air is applied to the induction system after the engine has reached operating temperature.

GENERAL MOTORS, CHRYSLER AND FORD UNIT

A quick check to determine if the automatic air temperature device is functioning properly is to look into the air cleaner snorkel before the engine is started. The damper door should be open so the air cleaner element should be visible. A mirror may be required depending on the air cleaner installation. If the snorkel passage is closed, trouble in the damper is indicated.

The vacuum-bleed sensor unit can also be easily tested. Remove the air cleaner cover and tape a small thermometer next to the sensor unit. The tape will prevent the thermometer from being sucked into the engine when it is started. Lay the cover back on the cleaner. With a cold engine, at idle speed, the damper door should be closed. (If the temperature is above 85° F. the damper door will be partially open and it will only be necessary to make certain that the damper is completely opened at 130° F.) Observe the action of the damper in the snorkel as the engine warms up. When the damper begins to open, lift the air cleaner cover and check the thermometer reading. At a temperature of about 85° F. the damper should start to open and at approximately 130° F. it should be fully opened.

Since the temperature sensor is preset and is nonadjustable, a malfunctioning unit must be replaced. It is held in place by two flat retaining clips. The new unit and gasket must be placed in the air cleaner housing in the same relative position as the original unit.

To check the operation of the vacuum motor, connect a test vacuum hose from the intake manifold to the motor. With the engine idling, the damper should close the cold air passage from the snorkel. A defective motor can be removed by drilling out the retaining spot welds on the General Motors air cleaner and securing the new motor in place with self-tapping sheet metal screws. The Chrysler vacuum diaphragm is removed by bending down the front lock tab, lifting the front edge of the motor, disengaging the rear tab, and then unhooking the operating rod from the heat control door.

FORD AND AMERICAN MOTORS UNIT

As an initial check on a cold engine with underhood ambient temperature of less than 100° F., look into the air cleaner snorkel. The valve plate should be in the heat-on or Up position. If the valve is not in the proper position remove the duct and valve assembly and submerge it in a container of cool water. With a thermometer placed in the water, slowly heat the water. At a water temperature of 100° F. or colder, the valve should be in the heat-on or Up position. If the valve does not move, check for possible mechanical interference of the plate in the duct.

Raise the water temperature to approximately 110° F., allow a few minutes for the temperature to stabilize, and the valve should start to open. Increase the water temperature to approximately 135° F. and again allow for temperature stabilization. The valve should now be in the heat-off or Down position. If the valve does not conform to these requirements, replace the unit with a new valve and duct assembly.

The override vacuum motor can be checked in the following manner. With a cold engine and underhood temperature at less than 100° F., start the engine and observe the position of the valve plate. It should be in the heat-on or Up position. If it is not, remove the hose from the vacuum motor and check the available vacuum at the hose. It should be at least 15 inches at idle speed. If the vacuum is less than 15 inches, check for vacuum leaks in the hose and hose connections. If the proper amount of vacuum is applied to the motor but the valve plate is not in the heat-on position, the motor is defective and should be replaced.

The Ford Auxiliary Air Inlet Valve should be fully closed with the engine idling. Disconnecting the motor vacuum hose should permit the valve to open fully. To check for faulty operation, check for interference caused by misalignment of the valve plate or motor rod. Then test the vacuum applied to the motor. There should be a minimum of 15 inches. If the vacuum is below this figure, check for hose and connection leaks. If vacuum is sufficient but the valve remains in one position, remove the vacuum motor from the air cleaner housing and test its operation at any other vacuum source. Replace the motor if it is defective.

SERVICE TIP

When removing the heated carburetor air cleaner from the engine, it is important to remember that a vacuum hose is connected to the underside of the cleaner housing. The hose from the Closed PCV System may also be attached to the air cleaner housing. It is essential, of course, that these hoses be reconnected when the air cleaner is replaced on the carburetor.

Be careful not to damage the plastic fittings which are frequently used in connection with vacuum hoses.

CARBURETOR ADJUSTMENT
Exhaust Emission Controlled Engines

CARBURETOR ADJUSTMENT

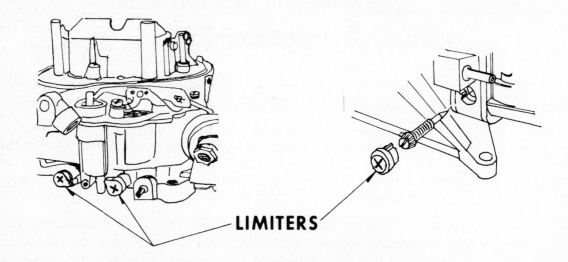

LIMITERS

EXTERNAL TYPE LIMITERS

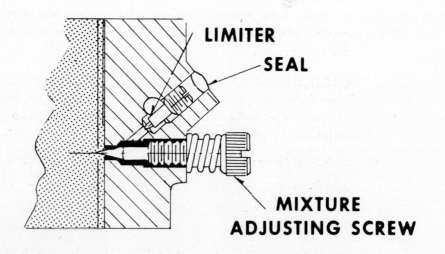

LIMITER

SEAL

MIXTURE
ADJUSTING SCREW

INTERNAL TYPE LIMITER

CARBURETOR ADJUSTMENT

EXHAUST EMISSION CONTROLLED ENGINES
As previously stated, the carburetors on emission controlled engines are calibrated for lean mixtures. By limiting the fuel mixture richness, the objectionable exhaust hydrocarbon emissions can be proportionately reduced.

IDLE LIMITERS
To prevent the idle mixture adjusting screws from being accidently or deliberately set too rich, many carburetors are presently being fitted with a device called an idle limiter.

Idle mixture adjustment limiters are of two types - external and internal. That is, they are externally mounted on, or internally positioned in, the carburetor.

The external type idle limiter is a plastic cap which fits over the head of the adjusting screw. The cap has a lug or stop projection which bumps the carburetor housing at each end of its travel thereby limiting the amount of adjustment possible. Two and four barrel carburetors have a limiter on each mixture adjusting screw.

The internal type limiter is located in a calibrated passage in the carburetor fuel idle circuit. The idle mixture is adjusted by turning the adjusting screw in the normal manner. But if the screw is backed out too far in an effort to enrichen the mixture excessively, the overadjustment has no effect on the mixture setting after a certain point, since the richness of the mixture is then controlled by the size of the limiter passage.

It is very important that the idle limiters are not removed from the carburetor or mutilated to increase the adjustment range. If all the other engine components have been properly tuned prior to carburetor adjustment, a satisfactory mixture and idle adjustment can be effected within the range provided by the limiters.

CARBURETOR "LEAN ROLL" ADJUSTMENT
The "lean roll" method of setting the carburetor is presently a recommended method of adjustment. Basically, this adjustment is performed in the following manner:

- Adjust the idle mixture screws, alternately, to obtain the highest idle rpm.

- Adjust the idle rpm to the specified speed. If the carburetor is equipped with an idle stop solenoid, adjust the idle speed with the solenoid adjusting screw.

- Turn one mixture screw clockwise, slowly, and in small increments, until the idle speed drops about 15 rpm, or as recommended. Turn the other mixture to obtain another 15 rpm drop.

- Readjust the slow idle speed to specifications.

  If the carburetor is equipped with an idle stop solenoid, disconnect the solenoid lead and set the low (shutdown) idle speed with the carburetor idle adjustment screw. Reconnect the solenoid lead.

Note that the final mixture adjustment is performed by turning the mixture screw clockwise (leaner) to effect the "lean roll" setting that permits the leanest air/fuel ratio consistent with effective control of emissions while providing acceptable engine performance.

Some carburetor mixture and idle adjustment specifications are listed as Initial and Final rpm settings. For example: Turn mixture screws IN until lightly seated, then back screws OUT 4 turns. Adjust carburetor idle speed screw (or idle solenoid screw) to obtain Initial idle speed (say 775 rpm). Then turn mixture adjusting screws IN, equally in small increments, until Final idle speed is obtained (say 700 rpm). Electrically disconnect solenoid, if so equipped, and adjust carburetor idle speed screw to obtain Shutdown idle rpm specified (say 400 rpm). This new procedure sets the idle speed by adjusting the mixture screws. Following the car makers recommended carburetor adjustment procedure "to the letter" is the best assurance that acceptable performance and effective emission control will both be achieved by the tune-up.

The correct carburetor adjustment procedure (along with the ignition timing and idle speed setting), is so important to effective emission control that starting with the 1968 models the car manufacturers have mounted a decal in each engine compartment listing these important specifications and outlining the carburetor adjustment procedure. Make it a practice to always follow these important recommendations.

It is also important that all special instructions be observed when adjusting the carburetor, as: headlights on high beam; air conditioning unit turned ON or OFF; hot idle compensator valve held closed; air cleaner ON or OFF and/or automatic transmission in Neutral or Drive. Always an important consideration is that the engine must be up to normal operating temperature before carburetor adjustment is performed.

The use of a combustion analyzer to measure the carburetor air/fuel ratio is presently being recommended by car manufacturers. Be sure to follow the equipment manufacturer's instructions when using the analyzer to secure accurate readings. It is important the carburetor mixture screws be turned no more than 1/16 turn at a time with a 10 second wait between adjustments. It normally takes this long for the meter to sense the change in mixture setting.

The use of the Exhaust Gas Analyzer as a testing device is increasing in popularity. The analyzer measures the concentrations of hydrocarbons (HC) and carbon monoxide (CO) emissions in the exhaust gases. The hydrocarbon reading is a measurement of the unburned fuel, usually in parts per million (ppm), in the exhaust gases. The carbon monoxide reading is a relative measurement of combustion efficiency.

Like other specialized test equipment, the Exhaust Gas Analyzer assists the tune-up technician in performing professional service by directing him to the possible causes of trouble. High hydrocarbon (HC) readings indicate such conditions as: misfiring or fouled spark plugs, overadvanced ignition timing, defective breaker points, vacuum leaks or disconnected vacuum lines. High carbon monoxide (CO) readings indicate: carburetor misadjustment, rich choke setting or defective choke action, high float level, restricted or plugged PCV valve, or a dirty carburetor air cleaner. And if both hydrocarbon and carbon monoxide readings are high, both the ignition and the carburetion systems need attention which usually means a major tune-up should be performed. Used both before and after a tune-up, the Exhaust Gas Analyzer will reveal the presence of trouble and verify its correction.

As exhaust emission standards become mandatory throughout the country, the Exhaust Gas Analyzer will become an essential piece of test equipment.

It is important to have your instrument tested for accuracy and recalibrated as required at specified intervals to insure accurate readings.

THE VITAL IMPORTANCE OF PRECISION TUNE-UP ON EMISSION CONTROLLED ENGINES

The various systems of automotive emission control are not difficult to maintain. There are, however, certain critically essential requirements.

Major tune-up **must** be performed at the time or mileage interval recommended by the vehicle manufacturer. The engines must be **precisely** tuned to the car maker's specifications with the use of **quality** test equipment. Never before has **precision** tuning been such an absolute requisite.

Tune-up now serves two important functions - it must effectively control engine emissions - and it must keep the car owner pleased with over-all engine performance. Only a professional precision tune-up can satisfy both requirements.

A very important point to remember when tuning emission controlled-engines is that there is no single unit or device that is the major controlling factor in tuning the engine. Every engine accessory or assist unit is designed to function in relation to other units. It is important that every step in the tune-up procedure be meticulously performed. There are no short cuts - no minor tune-ups. Even minor diviations from some specifications can have undesirable effects on emission control.

An important fact for the tune-up technician to remember is that even after a tune-up the modern emission-controlled engine may not have the performance capability and the smooth idle of former models. The reason is that air/fuel mixtures and ignition timing have been altered considerably to minimize air pollutants from the automobile engine. This minor inconvenience is very small payment for the cleaning of the air, and is an effort in which ALL motorists must participate as more states pass, or contemplate passing, "Clean Air" legislation. It is particularly important that the carburetor be precision adjusted to limit exhaust emissions rather than to effect a smooth idle.

CARBURETOR

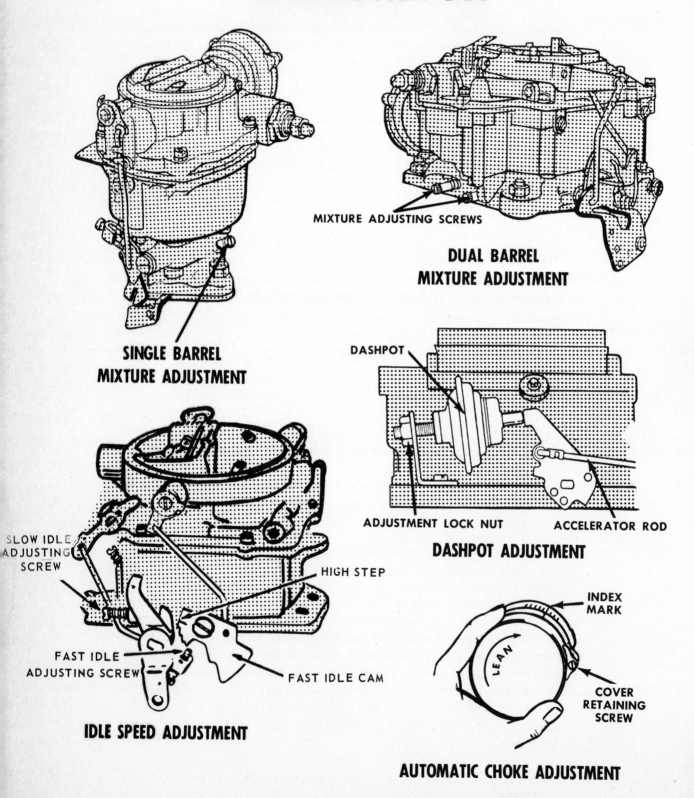

SINGLE BARREL
MIXTURE ADJUSTMENT

MIXTURE ADJUSTING SCREWS

DUAL BARREL
MIXTURE ADJUSTMENT

DASHPOT

ADJUSTMENT LOCK NUT   ACCELERATOR ROD

DASHPOT ADJUSTMENT

SLOW IDLE
ADJUSTING
SCREW

HIGH STEP

FAST IDLE
ADJUSTING SCREW

FAST IDLE CAM

IDLE SPEED ADJUSTMENT

INDEX
MARK

LEAN

COVER
RETAINING
SCREW

AUTOMATIC CHOKE ADJUSTMENT

CARBURETOR ADJUSTMENT

NONEXHAUST EMISSION CONTROLLED ENGINES

The adjustment of the carburetor is the last step in the tune-up procedure. It is performed after all the other tests and adjustments that influence engine operation have been completed. The carburetor is adjusted only with the engine at normal operating temperature.

The general carburetor adjustment procedure is as follows:

- Connect a tachometer and/or a vacuum gauge to the engine.

- Start and idle the engine until normal operating temperature is reached. Be sure the engine is running on slow idle.

- Turn the idle mixture adjusting screw in slowly until tachometer or vacuum gauge needle drops slightly.

- Turn idle mixture adjusting screw out until tachometer or vacuum gauge returns to highest reading.

- Repeat procedure on the other idle adjusting screw (if the carburetor is so equipped). Go back to the first adjusting screw and "trim off" the adjustment.

- Adjust the throttle stop screw so that the engine idles at the specified rpm. Turn on the headlights or the air conditioning unit when making this adjustment, if so instructed.

- Set fast idle adjustment to specifications.

After the carburetor is adjusted, operate the throttle linkage to check for binding and for accelerating pump discharge.

If the carburetor is equipped with a dashpot for slow throttle closing, it may be checked by allowing the throttle to snap shut several times while observing the tachometer after each check. The engine should return to the same idle speed each time. If it does not, the linkage may be sticking or the dashpot may be malfunctioning. Relieve the binding in the linkage and replace the dashpot if it does not respond to adjustment.

Hand-operate the choke linkage for free operation. If sticking or binding exists it may be caused by gum or varnish on the choke shaft or by carbon on the choke piston. The condition may be corrected by removing the carburetor air cleaner and the choke cover and flushing the choke shaft bearings and the choke piston with solvent while hand-operating the choke linkage. If the condition is not relieved, the carburetor should be removed from the engine for complete cleaning. When reinstalling the choke cover be sure it is positioned properly in reference to the index adjusting mark (e.g.: 1 rich; 2 lean).

Carburetors that do not respond to adjustment must be removed from the engine for disassembly and complete cleaning.

CHARGING VOLTAGE TEST

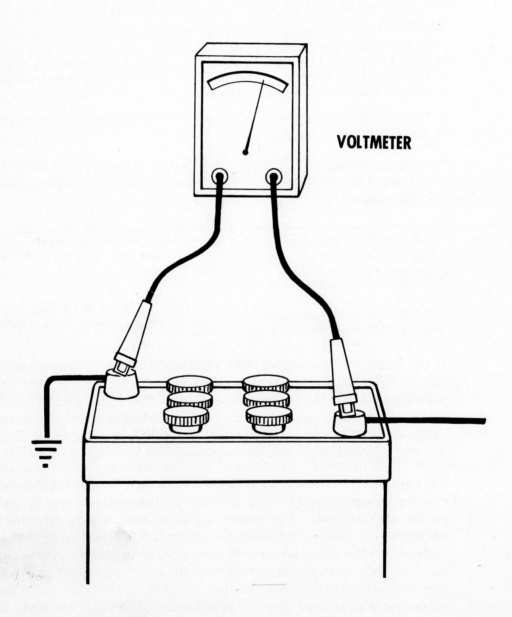

VOLTMETER

CHARGING VOLTAGE TEST

The charging system voltage test is another of the area tests. This test is a good indication of the overall operation of the electrical system including generator output and voltage regulator setting. The test also indicates the voltage applied to the ignition system which is an important factor in cases of breaker point burning and when short operating life of other electrical units is experienced.

To conduct the charging voltage test, the voltmeter is connected across the battery. Operate the engine at a fixed speed of 1500 to 1800 rpm. Observe the voltmeter for the charging voltage reading. If there is a tendency for the voltage to climb slightly, wait for the highest reading. Compare the highest voltage reading to specifications.

The charging voltage specification for a 12-volt system is generally between 13.8 and 15.0 volts. The specifications for a 6-volt system is generally between 6.8 and 7.5 volts.

Voltage readings higher than specified indicate a defective or misadjusted voltage regulator, high resistance in the regulator ground circuit, or a defective generator field circuit.

Voltage readings lower than specified indicate a loose drive belt, a defective generator, a defective or misadjusted voltage regulator, high resistance in the charging circuit or a discharged battery.

It is assumed that a partially charged battery, which would cause a low false charging voltage test reading, would have been revealed in the cranking voltage test, the first test in the procedure. However, if it is suspected that an undercharged battery is causing the low voltage reading, the battery can be quickly checked as being the cause of the trouble by the following method. Connect a test ammeter at the battery terminal of the regulator. If, during the voltage test, more than 10 amperes are flowing, the battery is too low to make an accurate voltage test. To proceed in properly conducting the charging voltage test, either place the car battery on fast charge; replace the car battery with a fully charged battery to complete the test; or connect a ¼ ohm resistor in series with the battery to simulate a fully charged battery condition.

Voltage readings either below or above the specified setting indicate the need for a complete charging system test.

MERCHANDISING TUNE-UP SERVICE

- Driveway banners
- Window signs
- Emphasize "Clean Air" Emission Control Center
- Display tune-up test equipment
- Display worn-out or damaged ignition parts
- Attractive ignition parts stock display
- Display Step-By-Step Tune-Up Procedure poster
- Display diplomas and certificates
- Emphasize new car Warranty Tune-Up Service
- Conduct seasonal tune-up campaigns
- Conduct tune-up "Quick Check" service clinics
- Maintain follow-up records

MERCHANDISING TUNE-UP SERVICE

Tune-up is a popular, essential and profitable automotive service. Besides being profitable in itself, tune-up offers the profit opportunities associated with the sale of spark plugs, cables, breaker points, condensers, distributor caps and rotors, coils, batteries, regulators, fan belts, carburetors, fuel pumps, fuel filters and many other essential items. In fact, few other automotive services have the profit potential of tune-up.

The important factor in selling tune-up, is to explain the benefits of the service to the car owner. Many motorists haven't any idea, or only a vague idea, of what a tune-up really consists or the benefits they stand to realize from tune-up. When you explain the savings in fuel and the increased dependability and the driving pleasure he will receive from an expertly tuned engine, your tune-up service will be easy to sell.

There are several aids that will assist you in merchandising your tune-up service:

- Pump island and driveway banners advertising your professional tune-up service.

- Window signs advertising your tune-up service is performed by specialists using only quality replacement parts and serving as the "Clean Air" Emission Control Center in your area.

- An attractive display of tune-up test equipment.

- A display of worn out or damaged ignition parts.

- An attractive ignition parts cabinet or a neat arrangement of ignition parts on the shelves in your tune-up bay.

- Display your Step-by-Step Tune-Up Procedure Poster.

- Display your diplomas and certificates of graduation from trades schools and technical clinics that you have attended.

- Emphasize your part in maintaining new car warranty.

- Promote, advertise and conduct seasonal tune-up campaigns.

- Conduct Tune-Up "Quick Check" Service Clinics.

- Maintain follow-up records on tune-ups performed for future solicitations.

Your major objective in merchandising your tune-up service is to let every motorist know that you have the knowledge, training, equipment and organization to expertly perform this vital automotive service.

Banners and window signs are an excellent manner of constantly informing the car owners in your area that you are prepared to offer this service, with complete technical skill, the proper test equipment and the use of only precision-built, quality replacement parts.

With the continuing emphasis on the air pollution problem, your Emission Control Center status will further serve to emphasize your contribution to the "Cleaner Air" movement, through professional automotive tune-up.

An attractive display of your test equipment is always an attention-getter. The sight of instruments and meters have a fascination for most people and the display of your instruments will quickly bring them around to the subject of tune-up.

Samples of a few worn out or damaged units that you have taken from engines you have tuned may be worth a thousand words when discussing the necessity for tune-up with a prospect. Most car owners, particularly women motorists, have never seen a set of pitted breaker points or a fouled spark plug and they don't understand how a cracked distributor cap or an eroded distributor rotor can be the cause of their trouble.

An attractive ignition parts display cabinet or a neat shelf arrangement of ignition parts always creates a favorable impression. It impresses your customers with the fact that the neatness of your parts display will be reflected in the care and methodical thought you use when working on their engines.

Display your Step-by-Step Tune-Up Procedure Poster in your tune-up bay where your customers can see the many services you perform during your tune-up. The poster will also serve as an excellent aid in explaining the sequence of tests in the tune-up operation to the car owner to impress him with the thoroughness of your specialized service.

A display of your diplomas and certificates that you have earned by graduating from trade schools and technical clinics also favorably influence your customers. Honest and sincere dedication to the pursuit of knowledge in a chosen profession is the finest attribute that any man can possess.

Emphasize new car warranty service by informing your clients driving new cars of the importance of maintaining their cars in proper condition with your professional tune-up service to keep their cars in warranty.

Conducting seasonal campaigns is also an excellent manner of merchandising your tune-up service. After the rigors of winter car operation, tune-up is readily accepted as a major part of the spring and vacation "make-ready" service. Tune-up is also an essential part of every program to winterize the car in preparation for trouble-free, cold-weather driving.

These seasonal campaigns may be merchandised, for example as a "package promotion" incorporating an engine tune-up with chassis lubrication, motor oil change, oil filter change and brake adjustment. Automatic transmission service, front end alignment, tire rotation or front wheel bearing repack and adjustment are services that can also be included in various "packages" at different prices.

Conducting a tune-up "Quick Check" service clinic is an excellent way of finding the need for complete tune-up. Most cars are being driven with fouled spark plugs, incorrect breaker point dwell, improper ignition timing, misadjusted carburetors and other malfunctions. Conducting a service clinic has in most cases proven to be a successful and profitable merchandising effort.

The clinic may be conducted for one day or for several days. It should be promoted throughout the area for several days before its opening by printed circulars, newspaper ads or handbills.

The inducement of the clinic is usually to offer a minor tune-up inspection for a nominal charge, probably $2.98. The charge is low enough to attract the average car owner and yet is sufficient to absorb a portion of the inspection costs.

No parts are replaced for this charge. As the tune-up procedure is followed, the defects found are recorded and reported to the car owner and permission to perform the needed service in the future is solicited. Such operations as adjusting the point dwell, setting the ignition timing and adjusting the carburetor are performed as required. The customer is advised, however, that without the replacement of the defective parts, the full benefits of tune-up will not be realized.

In preparing for a clinic, conduct several practice runs on different makes of cars, running through the tune-up procedure step-by-step, with your test equipment until you are certain of the sequence of steps, the proper use of the equipment, and the time interval involved. The knowledge you develop will assure you of conducting an expert and convincing demonstration and permit you to properly pace yourself for the number of cars you intend to inspect in the time you have allotted.

Maintaining a follow-up system is an excellent way of keeping record of the time or mileage when another tune-up should be solicited. By follow-up solicitations you can control the flow of work through your shop for smoothest operation and maximum profits.

Do not overlook the opportunity of participating in neighborhood or area Safety Inspection Campaigns. These drives offer another excellent chance to keep your business developing and growing by providing this important service to the car owners in your community.

Above all, constantly solicit tune-up service. A follow-up system of cars on which you performed tune-up will remind you when the service is again due. Remember to check the doorjamb stickers at every opportunity, too.

When soliciting a tune-up remember to:
- Listen to the customer's complaint. It may give you a lead that will save you troubleshooting time. It may also alert you to the fact that the complaint cannot be corrected by tune-up. If the customer is not so informed he will unfairly criticize your tune-up effort.

- Follow your Ignition Manufacturers Institute tune-up procedure on every job. It is your assurance of being right the first time because it serves as a guide to lead you through a specialized tune-up, step-by-step.

- Schedule the job so that it fits into the tune-up bay work load smoothly during the hours of the day when shop time is available.